PUBLIC RESPONSES TO FOSSIL FUEL EXPORT

PUBLIC RESPONSES TO FOSSIL FUEL EXPORT

Exporting Energy and Emissions in a Time of Transition

Edited by

HILARY BOUDET

Sociology, School of Public Policy,
Oregon State University,
Corvallis, OR, United States

SHAWN HAZBOUN

Graduate Program on the Environment,
The Evergreen State College,
Olympia, WA, United States

Elsevier
Radarweg 29, PO Box 211, 1000 AE Amsterdam, Netherlands
The Boulevard, Langford Lane, Kidlington, Oxford OX5 1GB, United Kingdom
50 Hampshire Street, 5th Floor, Cambridge, MA 02139, United States

Library of Congress Cataloging-in-Publication Data
A catalog record for this book is available from the Library of Congress

British Library Cataloguing-in-Publication Data
A catalogue record for this book is available from the British Library

ISBN: 978-0-12-824046-5

For information on all Elsevier publications visit our website at https://www.elsevier.com/books-and-journals

Publisher: Joseph P. Hayton
Acquisitions Editor: Graham Nisbet
Editorial Project Manager: Aleksandra Packowska
Production Project Manager: Sreejith Viswanathan
Cover Designer: Greg Harris

Typeset by TNQ Technologies

For Jeff, who steadfastly supports my research while conducting his own. Thank you for the meals and hugs, the hours on weekends to think and write, and for truly doing your share of the parenting (and holding the baby as I type this). —Shawn Hazboun

For the Boudet boys, who make sure that I don't get too wrapped up in work and give me hope for the future. —Hilary Boudet

Contents

PART III Public opinion on export

Contributors

Gisle Andersen
Norwegian Research Centre (NORCE), Social Science Department, Bergen, Norway; University of Bergen, Centre for Climate and Energy Transformation (CET), Bergen, Norway

Clifford Gordon Atleo (Niis Na'yaa/Kam'ayaam/Chachim'multhnii)
School of Resource and Environmental Management, Simon Fraser University, Burnaby, BC, Canada

Claudia F. Benham
School of Earth and Environmental Sciences, The University of Queensland, St Lucia, QLD, Australia

Emily Paige Bishop
School of Environmental Studies, University of Victoria, Victoria, BC, Canada

Hilary Boudet
Sociology, School of Public Policy, Oregon State University, Corvallis, OR, United States

Tyla Crowe
School of Resource and Environmental Management, Simon Fraser University, Burnaby, BC, Canada

Peter Erickson
Stockholm Environment Institute US Center, Seattle, WA, United States

Farid Guliyev
Department of Political Science and Philosophy, Khazar University, Baku, Azerbaijan

Shawn Hazboun
Graduate Program on the Environment, The Evergreen State College, Olympia, WA, United States

Anne N. Junod
Urban Institute, Washington, DC, United States

Tamara Krawchenko
School of Public Administration, University of Victoria, Victoria, BC, Canada

Julia Loginova
School of Earth and Environmental Sciences, The University of Queensland, St Lucia, QLD, Australia

Huy Nguyen
Computer Sciences Program, The Evergreen State College, Olympia, WA, United States

Åsta Dyrnes Nordø
Norwegian Research Centre (NORCE), Social Science Department, Bergen, Norway

Jonathan Pierce
University of Colorado Denver, School of Public Affairs, Denver, CO, United States

Georgia Piggot
School of Environment, University of Auckland, Auckland, New Zealand; Stockholm Environment Institute US Center, Seattle, WA, United States

Kathleen Saul
Graduate Program on the Environment, The Evergreen State College, Olympia, WA, United States

Karena Shaw
School of Environmental Studies, University of Victoria, Victoria, BC, Canada

Greg Stelmach
Sociology, School of Public Policy, Oregon State University, Corvallis, OR, United States

Endre Meyer Tvinnereim
University of Bergen, Department of Administration and Organization Theory, Bergen, Norway; University of Bergen, Centre for Climate and Energy Transformation (CET), Bergen, Norway

Richard Weiss
Computer Sciences Program, The Evergreen State College, Olympia, WA, United States

Patricia Widener
Sociology, Florida Atlantic University, Boca Raton, FL, United States

Chad Zanocco
Civil and Environmental Engineering, Stanford University, Stanford, CA, United States

Acknowledgments

We first wish to thank all the authors who contributed to this work—collectively, your research has begun to close the knowledge gap on the social dimensions of fossil fuel export. We would also like to thank the research participants who provided the data for the empirical chapters in this volume. We know that energy communities often get repeated requests to participate in social science research and that this can be draining—your input is invaluable, thank you. Last, we would like to acknowledge the presenters and audience members of the "Public responses to shifting fossil fuel regimes" session at the 2nd International Conference on Energy Research & Social Science (ERSS 2019)—this volume was born out of that session.

PART I

Introduction

CHAPTER 1

An introduction to the social dimensions of fossil fuel export in an era of energy transition

Hilary Boudet[1] and Shawn Hazboun[2]
[1]Sociology, School of Public Policy, Oregon State University, Corvallis, OR, United States; [2]Graduate Program on the Environment, The Evergreen State College, Olympia, WA, United States

The twin forces of the climate crisis and the shale revolution have created profound changes in global energy markets. Mounting pressures to cut carbon emissions have accelerated national policies to facilitate transitions away from coal and oil to natural gas and ultimately renewable energy sources. At the same time, technological advances combining horizontal directional drilling and hydraulic fracturing have opened vast previously infeasible reserves of oil and gas to development. Moreover, the COVID-19 pandemic has had vast impacts on human behavior and mobility around the world, affecting energy and electricity markets in unprecedented ways that will likely persist beyond the health crisis. The result has been changes in the energy markets and trade that would have seemed unfathomable just two decades ago. For example, the United States, which was facing dire predictions of natural gas shortages in the 1990s, prompting a flurry of proposals for import facilities, became a net energy exporter for the first time in 70 years in 2019 (U. S. Energy Information Administration, 2021). Oil production and exportation in North America has also broken OPEC's stronghold on oil prices. Moreover, in Australia, despite dwindling domestic consumption, coal became its most valuable export in 2018, with most of its production going to East Asia.

Yet, fossil fuel export has not come without controversy. Increasingly, environmental activists focused on keeping fossil fuels in the ground to prevent their combustion from contributing to climate change have targeted not just extractive sites, but the pipelines, railways, and export terminals meant to facilitate the export of these fuels overseas. Despite its importance to global energy markets, research on public perceptions and responses to fossil fuel export has received relatively scant attention from social scientists—who have preferred to focus on extraction and production

Public Responses to Fossil Fuel Export
ISBN 978-0-12-824046-5
https://doi.org/10.1016/B978-0-12-824046-5.00014-X

sites. Here, we bring together a volume exclusively about fossil fuel export, covering social science research from a variety of disciplines about a variety of fuel types.

Our aim is to provide wide-ranging perspectives—both theoretically and methodologically—on the human dimensions of fossil fuel export. What do members of the public think about exporting fossil fuels in places where it is happening? What do they see as its main risks and benefits? What connections are being made to climate change and the impending energy transition? How have those communities most affected responded to proposals related to fossil fuel export, broadly defined to include transport by rail, pipeline, ship, etc.? These are the research questions that underpin this book.

In the introduction to this volume, we begin with a brief overview of the status of fossil fuel export around the world, including an overview of recent trends and the impact of the COVID-19 pandemic. We then provide a review of relevant literature, with a particular focus on studies of public perceptions and community response to energy technologies. We conclude with information about the organization of the book.

Fossil fuel export: status and trends

The COVID-19 pandemic has had and continues to have a significant impact on economies around the world. The energy sector—which is fundamentally tied to economic activity—was not immune. According to the International Energy Agency (2021), global energy demand fell 4% in 2020—the largest drop since World War II. Oil demand, which fuels much of global transport, was particularly hard hit by travel restrictions implemented to slow the spread of the virus. Yet, 2021 has brought stimulus packages, vaccines, and economic recovery to some parts of the world, which have resulted in a corresponding growth in energy demand—particularly in China where early and aggressive containment of the virus has allowed life to largely return to normal with associated increases in demands for energy services. In fact, global energy demand is expected to grow 4.6% in 2021, thus counteracting 2020's contraction, with 70% of this increase expected to come from emerging markets and developing economies.

Unfortunately, this increased energy demand will result in increased carbon emissions—energy-related carbon emissions are expected to experience their second biggest increase ever in 2021, reversing 80% of the decline experienced in 2020. Increased demand for all fossil fuels, in particular coal, is driving these increased emissions. In fact, coal demand is expected to exceed 2019 levels, largely driven by demands for power generation in Asia. Yet, demand for transport oil, specifically for aviation, will remain below 2019 levels, so emissions could have been higher with recoveries in all industries to prepandemic levels. Among fossil fuels, natural gas has been the most resilient to the pandemic's impacts, and demand is on track to have the largest increase compared to 2019 levels, growing 3.2% in 2021. This resilience has in part been driven by fuel switching from coal to natural gas for electricity generation (International Energy Agency, 2021).

Turning to the power sector, electricity demand is projected to increase 4.5% in 2021—its fastest growth in over a decade (International Energy Agency, 2021). Again, most of this increase in demand comes from emerging markets and developing economies, with China responsible for half of global growth. The good news is that renewables, particularly their use in the electricity sector, have grown throughout the pandemic. Their share of electricity generation is expected to grow to almost 30% in 2021—the largest since the industrial revolution. Again, China leads and is expected to account for almost 50% of this increase (International Energy Agency, 2021).

What will this mean for fossil fuel trade and associated proposals for its transportation and export? After the historic disruption of global energy trade in 2020, 2021 has been marked by growth, but also growing pains. Much of this growth in demand has been met using excess supply and reserves, but these are dwindling. Current supply has not kept up with increasing demand; prices have soared; and thus far Europe and China have suffered the brunt of the problem (International Energy Agency's, 2021d; Bradsher, 2021). Such issues are unlikely to remain geographically isolated, however.

Much uncertainty remains. Despite recent surges in demand for fossil fuels, pandemic-related shocks combined with aggressive policies in many countries aimed at more sustainable recoveries may mean that fossil fuel demand, in particular for oil and coal, is unlikely to follow prepandemic trends and may even peak earlier than previously predicted (International Energy Agency, 2021b). The level of uncertainty—created by subsequent COVID-19 variants, political will to maintain policies aimed at sustainable recoveries, price fluctuations, etc—makes prediction difficult. What we do know is that communities and citizens will continue to face proposals for energy infrastructure—both its generation and transportation.

Our hope is that this volume sheds additional light on how the public perceives these proposals and how they respond, particularly in the context of fossil fuels but also with relevant insights for the siting of renewables. Such insights will prove invaluable both in the current moment and as we (hopefully) transition into a more sustainable energy system of the future.

What do we already know?

Boomtowns, risk perceptions, and overadaptation

A long and extensive literature on energy facility siting and its effects exists. Some of its initial beginnings can be traced back to work by Gilmore (1976), outlining the boomtown effects of energy development on a hypothetical rural town in Wyoming, and work by Slovic (1987) on public perceptions of hazards, including nuclear energy. The 1970s, 1980s, and early 1990s saw a flurry of academic work on the topic in social psychology and rural sociology—particularly as it related to the development of the nuclear industry in the United States and Yucca Mountain as a disposal site (Flynn et al., 1992; Flynn & Slovic, 1995; Krannich et al., 1991; Kunreuther et al., 1988, 1990; Riley et al., 1993; Slovic et al., 1991)—but also exploring public perceptions and community responses to oil and gas development (Brown et al., 1989; England & Albrecht, 1984; Freudenburg, 1992; Freudenburg & Gramling, 1992, 1994; Gramling & Freudenburg, 1992; Krannich et al., 1991; Krannich & Greider, 1984; Molotch, 1970; Molotch & Lester, 1975). Findings from these studies underscored the role of both individual demographic and community contextual factors in shaping public perceptions of energy development, as well as how aspects of the hazard itself can shape views. They also outlined the influential boomtown model of the impacts of energy development on communities, focusing on social disruptions related to service provision (e.g., for education, policing, etc.) (Brown et al., 1989; Jacquet & Kay, 2014), while also highlighting how communities dependent on a particular industry can become "overadapted" or even addicted to extractive development due to its economic benefits (Freudenburg, 1992; Freudenburg & Gramling, 1992). They outlined the typical opportunities and threats linked to these types of proposals (e.g., to physical environments; cultural, social, political/legal, economic, and psychological systems), with an eye toward the development of a more comprehensive process for social impact assessment (Gramling & Freudenburg, 1992).

In terms of fossil fuel infrastructure beyond extraction, liquefied natural gas (LNG) facilities did receive some attention during this timeframe, as several were proposed (and some built) in the United States and Europe to import natural gas from foreign sources (Kunreuther et al., 1983; Kunreuther & Lathrop, 1981; Kunreuther & Linnerooth, 1984). These studies highlighted the role of context and exogenous events in shaping public perceptions. For example, in a study of LNG siting in California in the 1970s, Kunreuther and Lathrop (1981) described how an oil tanker explosion in the Los Angeles harbor the day after the City Council allowed work to begin on an LNG terminal led to construction being suspended a week later.

Locally unwanted land uses and the environmental justice movement

A slowdown in siting of both nuclear power (in part driven by the Three Mile Island accident and cost concerns) and oil and gas development (related to the 1980s oil glut) in the United States also resulted in less attention on this topic in the academy. Attention shifted to other energy-related topics, in particular how to lower demand in the wake of the 1970s oil embargo and our energy system's environmental impacts as concerns about pollution increased and the environmental movement strengthened (Rosa et al., 1988).

During this same timeframe, a related set of literature was also growing in the planning and public policy fields exploring opposition to locally unwanted land uses (LULUs) more broadly (beyond energy), including incinerators, landfills, hazard waste sites, prisons, highways, etc (Dear, 1992; Futrell, 2003; Lesbirel, 1998; Schively, 2007; Smith & Marquez, 2000). Such facilities often create benefits to larger society but result in a set of risks to the local host community. At first, much of this literature focused on overcoming so-called NIMBYism, short for "Not in My Back Yard"—or an observed tendency for community members to oppose development near them (Inhaber, 1998; Rabe, 1994). As the field developed, however, scholars became less focused on overcoming NIMBYism and more focused on understanding it and even justifying the concerns of those most proximate to development (Boholm & Lofstedt, 2013; Hager & Haddad, 2015). The term NIMBY has now taken on a pejorative tone—casting opponents as self-interested, ignorant, and irrational—when opponents are often community-minded, well-informed, and rational (Dear, 1992; Rand & Hoen, 2017; Schively, 2007; Wolsink, 2006, 2007; Wüstenhagen et al., 2007).

Scholars have moved away from a focus on educating an uninformed public and toward more inclusive, participatory decision-making around the siting of unwanted facilities, including pushing for adequate compensation for those most affected (Arnstein, 1969; Devine-Wright, 2005, 2013; Kunreuther et al., 1987; O'Hare & Sanderson, 1993; Portney, 1984; Rabe, 1994).

What was becoming clear, thanks in large part to the environmental justice movement, was that LULUs and other hazardous facilities were overwhelmingly being sited in disadvantaged communities (Bullard, 1990; Mohai et al., 2009; Roberts & Toffolon-Weiss, 2001; Schlosberg, 2009; Taylor, 2014). Environmental justice scholars and activists began to highlight the ways in which segregation, zoning laws, and business practices allowed toxic facilities to be sited in racially segregated, low-income communities, exposing community members to toxic substances and creating health and other problems (Taylor, 2014). In studies of energy development, scholars have begun to refer to some areas as "energy sacrifice zones"—places that have been permanently damaged by extensive and continuous energy development, e.g., West Virginia's coal fields, coastal Louisiana's oil and gas fields (Bell, 2014; Lerner, 2012; Maldonado, 2018). Environmental justice concerns have also plagued fossil fuel transport infrastructure, e.g., the development of LNG terminals in Mexico to serve California markets (Carruthers, 2007). Scholars have offered economic, sociopolitical (i.e., "path of least resistance"), and racial discrimination explanations for these injustices, as well as principles for environmental justice (Harlan et al., 2015; Mohai et al., 2009). These principles, which have been carried forward into scholarship and movements related to energy and climate justice, include distributional, procedural, and recognitional justice (Harlan et al., 2015; Jenkins et al., 2016).

Renewables and fracking

Specific to energy facility siting, a growing literature began to focus on the siting of renewables. With proposals for wind energy beginning to take off, social scientists in both the United States and Europe began to study opposition. A "social gap" was observed between widespread support for renewables but local opposition to particular projects. Reviewing 30 years of literature on public acceptance of wind in North America, Rand and Hoen (2017) found that support for wind development has generally been high (higher than fossil fuels), with socioeconomic, visual, and sound impacts strongly tied to opposition and acceptance, more so even than environmental impacts (Haggett, 2011). Like previous studies of fossil fuel development and LULUs, perceptions of decision-making processes related to fairness, participation, and trust matter. Despite the social gap, NIMBYism has largely been debunked in these studies with proximity effects on support or opposition unclear (Wolsink, 2006, 2007). Instead, explanations for opposition have focused on place attachment—emotional connections to particular locations—and the place-protective actions residents are willing to take to prevent changes (Devine-Wright, 2009, 2011; Firestone et al., 2017).

More recently and perhaps most relevant to our subject matter, a flurry of academic activity has surrounded public perceptions and community responses to fracking. Fracking—or hydraulic fracturing—combined with horizontal directional drilling has opened vast resources in shale rock for extraction. This technological innovation is largely responsible for many of the changes to the energy market described in this book, as well as the increased push to export fossil fuels, particularly natural gas. Community response and public perceptions of fracking at the extraction site have received a lot of attention from scholars and in the media, particularly as the technology became increasingly controversial (Aczel & Makuch, 2018; Boudet et al., 2014, 2016, 2018; Alcorn et al., 2017; Bomberg, 2017; Brasier et al., 2011; Bugden et al., 2017; Bugden & Stedman, 2019; Cotton, 2013; Craig et al., 2019; Davis & Fisk, 2014; Dokshin, 2016, 2021; Evensen, 2018; Evensen et al., 2014, 2017; Graham et al., 2015; Jacquet et al., 2018; Jerolmack & Walker, 2018; Junod & Jacquet, 2019; Lachapelle et al., 2018; Ladd, 2014; Mayer, 2016; Thomas et al., 2017). Much of this literature has drawn on concepts and conceptual frameworks from the studies described previously—the boomtown model, risk perception studies, the role of context and place in shaping perceptions and response, opportunity-threat, environmental justice, etc. Other theoretical and conceptual frameworks have been added—including concepts from the study of social movements and social representations theory. Concepts from social movements' studies include framing, resource mobilization, and political opportunity structure, with additional focus on the strategies and tactics used by project supporters and opponents (Aldrich, 2008; Boudet, 2011; Boudet & Ortolano, 2010; Cheon & Urpelainen, 2018; McAdam & Boudet, 2012; McAdam et al., 2010; Sherman, 2011; Vasi et al., 2015; Walsh, 1981; Wright & Boudet, 2012).

In Social Representation Theory, social representations are "common sense understandings of complex, often scientific, phenomena, generated in the public sphere and reliant on the history, culture, and social structure of the context in which they emerge (Evensen, Clarke, & Stedman, 2014 - p. 63)" (Aczel & Makuch, 2018; Alcorn et al., 2017; Bomberg, 2017; Boudet et al., 2016, 2014, 2018; Brasier et al., 2011; Bugden et al., 2017; Bugden & Stedman, 2019; Cotton, 2013; Craig et al., 2019; Davis & Fisk, 2014; Dokshin, 2016, 2021; Evensen, 2018; Evensen, Jacquet, et al., 2014; Evensen et al., 2017; Graham et al., 2015; Jacquet et al., 2018; Jerolmack & Walker, 2018; Junod & Jacquet, 2019; Lachapelle et al., 2018; Ladd, 2014; Mayer, 2016; Thomas et al., 2017). Social Representations Theory—which has been applied to both fracking and renewable energy development—postulates that such representations are socially constructed and used to make the unfamiliar familiar (Batel & Devine-Wright, 2015; Bugden et al., 2017; Evensen & Stedman, 2016). In many ways, this theory echoes at the societal level many of the advances in human understanding and decision-making we now know at the individual level. People often have neither the time nor resources to fully consider the range of risks and benefits posed by a proposal for energy development. Instead, they rely on mental shortcuts (e.g., opinions of trusted individuals or elites, media, values, ideological predispositions, etc.) to decide their stance (Clarke et al., 2015; Ho et al., 2018, pp. 1–15; Jacquet, 2012; Vasi et al., 2015). These sources can provide social representations of a new technology—like fracking—as an economic boon or a controversial, unsafe technology (like the idea of framing prevalent in the social movement literature). These social representations shift over time and may differ by place. Indeed, research on media coverage of fracking in Europe and the United States has uncovered a similar pattern of competing discourses of economic opportunity versus environmental threat, yet with distinct contextual differences (Bomberg, 2017; Cantoni et al., 2018; Cotton, 2013; Dokshin, 2021; Jaspal et al., 2014). We might expect the same in coverage and debates on fossil fuel export to the extent such projects are connected to fracking via social representations—and indeed the few examples we have of this sort of scholarship on export infrastructure have found this to be the case (Chen, 2020; Chen & Gunster, 2016; Pierce et al., 2018; Tran et al., 2019).

Summary

Summarizing the literature on public perceptions of and community responses to new energy technologies, Boudet (2019) outlined four categories of relevant factors: technology, people, place, and process (see Fig. 1.1). We see these factors woven throughout the literature described above. For technology, risk and benefit perceptions, particularly in relation to alternative choices, matter. Moreover, these risk and benefit perceptions can

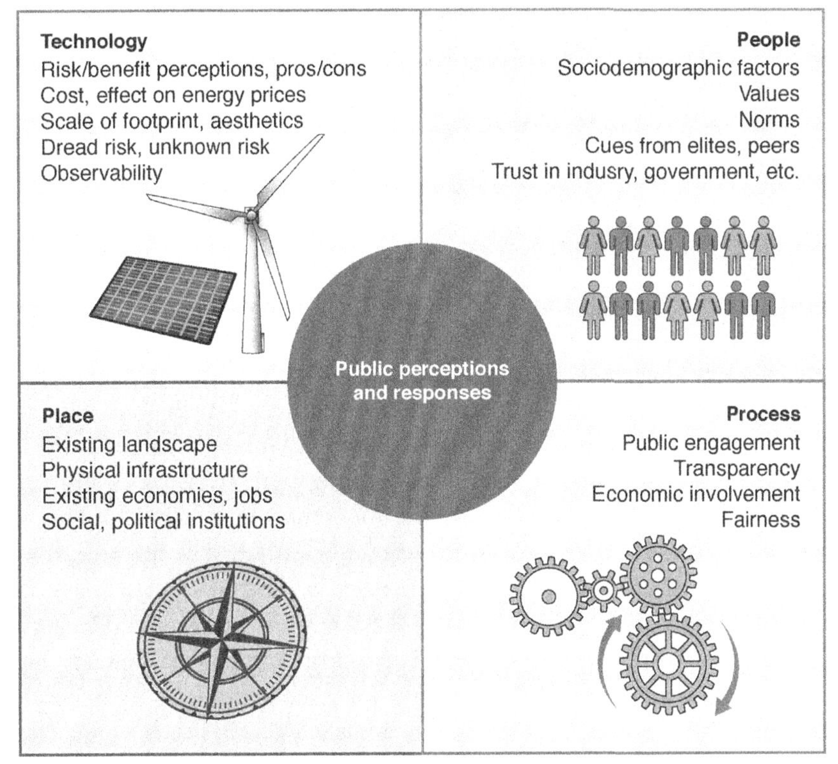

Figure 1.1 Factors affecting public perceptions of and responses to new energy technologies, reprinted from Boudet (2019). *(From Boudet, H. S. (2019). Public perceptions of and responses to new energy technologies. Nature Energy, 4(6), 446–455. https://doi.org/10.1038/s41560-019-0399-x)*

change over time—as has been happening in the case of natural gas. Once viewed as an environmentally friendly alternative to coal for electricity production, concerns about carbon emissions mean that the risks of increasing dependency on natural gas are being weighed more heavily, moving the discourse on this fuel from one that could serve as "a bridge to a renewable energy future" to one that instead could serve as "a bridge to nowhere" (Delborne et al., 2020; Hazboun & Boudet, 2021). Yet, aspects of the people involved also shape benefit and risk perceptions. Demographics—in particular, gender, race, and political ideology—have been shown to be important predictors on views of a whole host of technological risks, energy included. Dubbed the "conservative, white male effect," research has consistently shown this group to be more likely to accept new technologies and less concerned about associated environmental risks (Bergquist et al., 2020; McCright & Dunlap, 2013). Place, as highlighted above, can also shape views—proximity, rurality, experience with extractive industries, economic identity, presence of natural amenities, etc—can all shape how a new proposal is viewed (Boudet et al., 2016; Giordono et al., 2018; Hazboun & Boudet, 2020; Wright & Boudet, 2012). In fact, scholars have begun to call for research looking at the interaction between technology, people, and place to better delineate both geographies of perception (Zanocco et al., 2019, 2020) and impact (Haggerty et al., 2018) in energy development. And, finally, the decision-making process and how it is viewed by affected stakeholders can impact perceptions and response (Firestone et al., 2017; Wolsink, 2007).

While these factors are clearly important, researchers in this area tend to draw on many different theoretical frameworks in their work—from psychology, sociology, planning, public policy, geography, etc. Such diversity in approaches makes generalization difficult, but perhaps this is the nature of the issue. With context and place playing such an important role in perceptions and responses, it might be difficult to generalize beyond typologies, like those incorporated into the idea of geographies of perception and impact (Haggerty et al., 2018). Such differences may also stem from different disciplinary traditions and methodological approaches. Two methodologies have clearly dominated this research area: case studies and surveys. These two approaches have different epistemological and ontological traditions but can complement one another. For this reason, we have purposefully included examples of both types of research (and more) in this volume.

Organization of the book

To better understand public responses to fossil fuel export, we invited submissions from scholars exploring these issues around the world. We have divided these contributions into three sections.

In Section 2, two chapters provide a broad overview of the new landscape of fossil fuels technology, supply, and policy. Specifically, in Chapter 2, Guliyev provides an assessment of the new global energy order: a hybrid and shifting one in which the established power of the conventional fossil fuel industry dominated by a few large companies confronts an increasingly dynamic and growing renewable energy sector. With the transition to renewables already happening, he compares it to past energy transitions, which have often proven messy, protracted affairs with clear winners and losers. In Chapter 3, Piggot and Erickson explore in more depth how changing priorities and the need to address the climate crisis are affecting fossil fuel export specifically. Past national climate policies have often ignored fossil fuel export, instead focusing on where the fossil fuel is burned. Sentiments are changing—in large part due to pressure from nongovernmental groups aimed at "keeping it [fossil fuel] in the ground" —and countries are now beginning to consider policies that address the "supply side" of fossil fuels.

Section 3 explores public opinion on fossil fuel export in several different locations around the world—specifically the United States, Canada, Norway, and New Zealand—using both surveys and case studies. In Chapters 4 and 5, Zanocco et al. and Stelmach et al. focus on views in the United States on natural gas export. Zanocco et al. look at trends in public opinion from 2013 to 2017, finding increasing support for natural gas export over time. Stelmach et al. examine risk and benefit perceptions, as well as attitudes toward regulation, of natural gas export. In both chapters, gender and political ideology drive preferences. In Chapter 8, Hazboun and Boudet explore preferences for fossil fuel export (oil, natural gas, and coal) in the Pacific Northwest, covering both the United States and Canada. They find that, while majorities of respondents oppose coal and oil export, opposition to natural gas export is less widespread. Similar drivers are important in shaping preferences in the Pacific Northwest, including gender and political ideology, but also views on climate change and familiarity.

In Chapter 7, Andersen et al. report on Norwegian public opinion of fossil fuel export. Norway provides an interesting contrast, in that most production occurs offshore and out of sight, incomes are regulated to serve the Norwegian welfare state, and large portions of the population work in the sector. Oil and gas development there has been relatively uncontroversial until recently. With the climate crisis, public opinion is changing. In Chapter 6, Widener's three case studies explore views toward export development in New Zealand, revealing similar tensions between the country's goals to achieve carbon neutrality and a focus on fossil fuel extraction for exportation.

The final section of the book contains six case studies that provide an in-depth look at how different communities around the world have been impacted—and responded to—fossil fuel export infrastructure. These include communities in Russia (Chapter 9), Canada (Chapters 10 and 11), Australia (Chapter 12), and the United States (Chapters 13 and 14). These case studies provide a deeper look at how export-related energy impacts might be similar to energy extraction- and production-related impacts, and themes arise that echo the findings of the energy impacts literature to date, including the importance of community participation (or the lack thereof) in decision-making and opposition, social disruption, concern about environmental impacts, justice, risk perception, and place attachment. Yet, these case studies also suggest that some impacts and perceptions are unique to fossil fuel export—for example, awareness of a community's role in the global fossil fuel supply chain and a perceived connection to other communities along transportation routes (pipelines, railways, etc.).

In Chapter 9, Loginova highlights the exclusionary processes that can undermine meaningful community participation in energy project decision-making using case studies of two communities in northern Russia. She notes four mechanisms by which exclusion occurs: discursive strategies, market mechanisms, legal and bureaucratic strategies, and strategies of uncertainty. In Chapter 10, Atleo et al. provide a critical examination of the varied (or "ambivalent") responses of 58 indigenous communities along the Trans Mountain Pipeline Expansion project in Canada. They posit that the history of colonialism in Canada has created a structure of uneven power relations that persists in contemporary dealings between Canada and First Nations, the latter of which must constantly assert their legal rights as sovereign nations. They propose a framework of six factors—rights, governance, economy, capacity, decision-making, and project—which influence whether a community will support or opposes the project. In Chapter 11, Bishop and Shaw examine how the community of Prince Rupert in British Columbia, Canada, responded to the Pacific Northwest LNG export terminal proposal. The authors find local opponents to be primarily worried about environmental impacts, specifically salmon runs. They also find similar themes of government distrust and dissatisfaction with the public engagement process, highlighted in multiple chapters in this section. In Chapter 12, Benham examines community impacts from LNG terminal development in two Australian locales. She highlights how terminal construction has been extremely rapid in these two remote communities, creating a boomtown effect. Benham finds significant social and psychological disruption from the development, findings that are consistent with decades of scholarly research on extraction-based boomtowns.

In Chapter 13, Junod conducts surveys in four railway communities to assess perceptions of oil transportation by railroad. Though oil by rail is an increasingly common method of transporting oil in the United States, its impacts on communities are understudied. Her findings highlight differences in risk perceptions based on community characteristics and experience with oil by rail accidents. In the final chapter of this section, Chapter 14, Hazboun et al. use computer-assisted content analysis of local news media to assess themes across two communities in the United States where large-scale coal export terminals were proposed. The authors draw on the concept of "scale shift" from social movement studies to explore the balance of local-scale themes with broader, "scaled up" themes, such as climate change and interstate commerce rights. They show that multiple nonlocal themes pervaded news coverage in both case studies, suggesting that the coal terminals engendered responses that went beyond community-level concerns.

Some concluding thoughts

Most current research focuses on public perceptions and community responses to fossil fuel extraction, such as hydraulic fracturing and coal mining, at the extraction site with little emphasis on the surrounding infrastructure required for such facilities to be economically viable. Yet, communities are responding in a variety of ways to this surrounding infrastructure—ways that are not adequately captured by current studies. Furthermore, as climate change awareness and activism has grown, more attention is being paid not only to phasing out domestic fossil fuel use but also to halting export of these fuels to other countries. We provide case examples and in-depth analysis of public opinion polls about fossil fuel export specifically.

The movement of fossil fuels across nations and around the world is a highly relevant component of the global energy system. The movement against fossil fuels has recognized the inherent vulnerability of such proposals and explicitly targeted them in their efforts to push for a more sustainable, carbon-free energy system. Given the globally mounting focus on the relationship between climate mitigation and energy policy, it is more vital than ever to understand the social forces influencing the global fossil fuel trade, including public opinion and community response toward fossil fuel infrastructure beyond the extraction site.

References

Aczel, M. R., & Makuch, K. E. (2018). *The lay of the land: The public, participation and policy in China's fracking frenzy*. The Extractive Industries and Society. https://doi.org/10.1016/j.exis.2018.08.001

Alcorn, J., Rupp, J., & Graham, J. D. (2017). Attitudes toward "fracking": Perceived and actual geographic proximity. *The Review of Policy Research, 34*(4). https://doi.org/10.1111/ropr.12234. n/a-n/a.

Aldrich, D. P. (2008). *Site fights: Divisive facilities and civil society in Japan and the west* (1st ed.). Cornell University Press http://www.jstor.org.stanford.idm.oclc.org/stable/10.7591/j.ctt7zc66.

Arnstein, S. R. (1969). A ladder of citizen participation. *Journal of the American Institute of Planners, 35*(4), 216–224.

Batel, S., & Devine-Wright, P. (2015). Towards a better understanding of people's responses to renewable energy technologies: Insights from Social Representations Theory. *Public Understanding of Science, 24*(3), 311–325. https://doi.org/10.1177/0963662513514165

Bell, S. E. (2014). "Sacrificed so others can live conveniently": Social inequality, environmental injustice, and the energy sacrifice zone of central appalachia. In C. M. Renzetti, & R. Kennedy-Bergen (Eds.), *Understanding diversity: Celebrating difference, challenging inequality*. Allyn and Bacon.

Bergquist, P., Konisky, D. M., & Kotcher, J. (2020). Energy policy and public opinion: Patterns, trends and future directions. *Progress in Energy, 2*(3). https://doi.org/10.1088/2516-1083/ab9592

Boholm, A., & Lofstedt, R. E. (2013). *Facility siting: Risk, power and identity in land use planning*. Routledge.

Bomberg, E. (2017). Shale we drill? Discourse dynamics in UK fracking debates. *Journal of Environmental Policy and Planning, 19*(1), 72–88. https://doi.org/10.1080/1523908X.2015.1053111

Boudet, H. S. (2011). From NIMBY to NIABY: Regional mobilization against liquefied natural gas in the United States. *Environmental Politics, 20*(6), 786–806.

Boudet, H. S. (2019). Public perceptions of and responses to new energy technologies. *Nature Energy, 4*(6), 446–455. https://doi.org/10.1038/s41560-019-0399-x

Boudet, H. S., Bugden, D., Zanocco, C., & Maibach, E. (2016). The effect of industry activities on public support for 'fracking. *Environmental Politics, 25*(4), 593–612. https://doi.org/10.1080/09644016.2016.1153771

Boudet, H. S., Clarke, C., Bugden, D., Maibach, E., Roser-Renouf, C., & Leiserowitz, A. (2014). "Fracking" controversy and communication: Using national survey data to understand public perceptions of hydraulic fracturing. *Energy Policy, 65*, 57–67. https://doi.org/10.1016/j.enpol.2013.10.017

Boudet, H. S., & Ortolano, L. (2010). A tale of two sitings: Contentious politics in liquefied natural gas facility siting in California. *Journal of Planning Education and Research, 30*(1), 5–21. https://doi.org/10.1177/0739456x10373079

Boudet, H. S., Zanocco, C. M., Howe, P. D., & Clarke, C. E. (2018). The effect of geographic proximity to unconventional oil and gas development on public support for hydraulic fracturing. *Risk Analysis, 38*(9). https://doi.org/10.1111/risa.12989

Bradsher, K. (2021). Power Outages Hit China, Threatening the Economy and Christmas. *New York Times*. https://www.nytimes.com/2021/09/27/business/economy/china-electricity.html. (Accessed 12 October 2021).

Brasier, K. J., Filteau, M. R., McLaughlin, D. K., Jacquet, J. B., Stedman, R. C., Kelsey, T. W., & Goetz, S. J. (2011). Residents' perceptions of community and environmental impacts from development of natural gas in the marcellus shale: A comparison of Pennsylvania and New York cases. *Journal of Rural Social Sciences, 26*(1), 32–61.

Brown, R. B., Geertsen, H. R., & Krannich, R. S. (1989). Community satisfaction and social integration in a boomtown: A longitudinal analysis. *Rural Sociology, 54*(4), 568. https://www.proquest.com/scholarly-journals/community-satisfaction-social-integration/docview/1290962572/se-2?accountid=14026.

Bugden, D., Evensen, D., & Stedman, R. (2017). A drill by any other name: Social representations, framing, and legacies of natural resource extraction in the fracking industry. *Energy Research & Social Science, 29*, 62–71. https://doi.org/10.1016/j.erss.2017.05.011

Bugden, D., & Stedman, R. (2019). Rural landowners, energy leasing, and patterns of risk and inequality in the shale gas industry. *Rural Sociology, 84*(3), 459–488. https://doi.org/10.1111/ruso.12236

Bullard, R. D. (1990). *Dumping in Dixie: Race, class, and environmental quality* (1st ed.). Westview.

Cantoni, R., Klaes, M. S., Lackerbauer, S. L., Foltyn, C., & Keller, R. (2018). Shale tales: Politics of knowledge and promises in Europe's shale gas discourses. *The Extractive Industries and Society, 5*(4), 535–546. https://doi.org/10.1016/j.exis.2018.09.004

Carruthers, D. V. (2007). Environmental justice and the politics of energy on the US–Mexico border. *Environmental Politics, 16*(3), 394–413. https://doi.org/10.1080/09644010701251649

Chen, S. (2020). Debating resource-driven development: A comparative analysis of media coverage on the Pacific Northwest LNG project in British Columbia. *Frontiers in Communication, 5*(66). https://doi.org/10.3389/fcomm.2020.00066

Chen, S., & Gunster, S. (2016). "Ethereal carbon": Legitimizing liquefied natural gas in British Columbia. *Environmental Communication, 10*(3), 305–321. https://doi.org/10.1080/17524032.2015.1133435

Cheon, A., & Urpelainen, J. (2018). *Activism and the fossil fuel industry*. Routledge. https://doi.org/10.4324/9781351173124

Clarke, C. E., Hart, P. S., Schuldt, J. P., Evensen, D. T. N., Boudet, H. S., Jacquet, J. B., & Stedman, R. C. (2015). Public opinion on energy development: The interplay of issue framing, top-of-mind associations, and political ideology. *Energy Policy, 81*, 131–140. https://doi.org/10.1016/j.enpol.2015.02.019

Cotton, M. (2013). Shale gas—community relations: NIMBY or not? Integrating social factors into shale gas community engagements. *Natural Gas & Electricity, 29*(9), 8–12. https://doi.org/doi:10.1002/gas.21678.

Craig, K., Evensen, D., & Van Der Horst, D. (2019). How distance influences dislike: Responses to proposed fracking in Fermanagh, Northern Ireland. *Moravian Geographical Reports, 27*(2), 92–107. https://doi.org/10.2478/mgr-2019-0008

Davis, C., & Fisk, J. M. (2014). Energy abundance or environmental worries? Analyzing public support for fracking in the United States. *The Review of Policy Research, 31*(1), 1–16. https://doi.org/10.1111/ropr.12048

Dear, M. (1992). Understanding and overcoming the NIMBY syndrome. *Journal of the American Planning Association, 58*(3), 288–300. https://doi.org/10.1080/01944369208975808

Delborne, J. A., Hasala, D., Wigner, A., & Kinchy, A. (2020). Dueling metaphors, fueling futures: "Bridge fuel" visions of coal and natural gas in the United States. *Energy Research & Social Science, 61*, 101350. https://doi.org/10.1016/j.erss.2019.101350

Devine-Wright, P. (2005). Beyond NIMBYism: Towards an integrated framework for understanding public perceptions of wind energy. *Wind Energy, 8*(2), 125–139. https://doi.org/10.1002/we.124

Devine-Wright, P. (2009). Rethinking NIMBYism: The role of place attachment and place identity in explaining place-protective action. *Journal of Community & Applied Social Psychology, 19*(6), 426–441. https://doi.org/doi:10.1002/casp.1004.

Devine-Wright, P. (2011). Place attachment and public acceptance of renewable energy: A tidal energy case study. *Journal of Environmental Psychology, 31*(4), 336–343. https://doi.org/doi.org/10.1016/j.jenvp.2011.07.001.

Devine-Wright, P. (2013). *Renewable energy and the public: From NIMBY to participation.* Routledge.

Dokshin, F. A. (2016). Whose backyard and whats at issue? Spatial and ideological dynamics of local opposition to fracking in New York state, 2010 to 2013. *American Sociological Review, 81*(5), 921–948. https://doi.org/10.1177/0003122416663929

Dokshin, F. A. (2021). Variation of public discourse about the impacts of fracking with geographic scale and proximity to proposed development. *Nature Energy*. https://doi.org/10.1038/s41560-021-00886-7

England, J. L., & Albrecht, S. L. (1984). Boomtowns and social disruption. *Rural Sociology, 49*(2), 230. https://www.proquest.com/scholarly-journals/boomtowns-social-disruption/docview/1290987822/se-2?accountid=14026.

Evensen, D. (2018). Yet more 'fracking' social science: An overview of unconventional hydrocarbon development globally. *The Extractive Industries and Society, 5*(4), 417–421. https://doi.org/10.1016/j.exis.2018.10.010

Evensen, D., Clarke, C., & Stedman, R. (2014a). A New York or Pennsylvania state of mind: Social representations in newspaper coverage of gas development in the marcellus shale. *Journal of Environmental and Social Sciences, 4*(1), 65–77. https://doi.org/10.1007/s13412-013-0153-9

Evensen, D., Jacquet, J. B., Clarke, C. E., & Stedman, R. C. (2014b). What's the "fracking" problem? One word can't say it all. *Extractive Industries and Society, 1*(2), 130–136. https://doi.org/10.1016/j.exis.2014.06.004

Evensen, D., & Stedman, R. (2016). Scale matters: Variation in perceptions of shale gas development across national, state, and local levels. *Energy Research & Social Science, 20*, 14–21. https://doi.org/10.1016/j.erss.2016.06.010

Evensen, D., Stedman, R., O'Hara, S., Humphrey, M., & Andersson-Hudson, J. (2017). Variation in beliefs about 'fracking' between the UK and US. *Environmental Research Letters, 12*(12), 124004. https://doi.org/10.1088/1748-9326/aa8f7e

Firestone, J., Hoen, B., Rand, J., Elliott, D., Hübner, G., & Pohl, J. (2017). Reconsidering barriers to wind power projects: Community engagement, developer transparency and place. *Journal of Environmental Policy and Planning, 20*(3), 370–386. https://doi.org/10.1080/1523908x.2017.1418656

Flynn, J., Burns, W., Mertz, C. K., & Slovic, P. (1992). Trust as a determinant of opposition to a high-level radioactive waste repository: Analysis of a structural model. *Risk Analysis, 12*(3), 417–429. https://doi.org/10.1111/j.1539-6924.1992.tb00694.x

Flynn, J., & Slovic, P. (1995). Yucca mountain: A crisis for policy: Prospects for America's high-level nuclear waste program. *Annual Review of Energy and the Environment, 20*(1), 83–118. https://doi.org/10.1146/annurev.eg.20.110195.000503

Freudenburg, W. R. (1992). Addictive economies: Extractive industries and vulnerable localities in a changing world Economy1. *Rural Sociology, 57*(3), 305–332. https://doi.org/10.1111/j.1549-0831.1992.tb00467.x

Freudenburg, W. R., & Gramling, R. (1992). Community impacts of technological change: Toward a longitudinal perspective*. *Social Forces, 70*(4), 937–955. https://doi.org/10.1093/sf/70.4.937

Freudenburg, W. R., & Gramling, R. (1994). *Oil in troubled waters: Perceptions, politics, and the battle over offshore drilling*. SUNY Press.

Futrell, R. (2003). Framing processes, cognitive liberation and NIMBY protest in the U.S. Chemical-weapons disposal conflict. *Sociological Inquiry, 73*(3), 359–386.

Gilmore, J. S. (1976). Boom towns may hinder energy resource development. *Science, 191*(4227), 535–540. http://www.jstor.org.stanford.idm.oclc.org/stable/1741302.

Giordono, L. S., Boudet, H. S., Karmazina, A., Taylor, C. L., & Steel, B. S. (2018). Opposition "overblown"? Community response to wind energy siting in the western United States. *Energy Research & Social Science, 43*, 119–131. https://doi.org/10.1016/j.erss.2018.05.016

Graham, J. D., Rupp, J. A., & Schenk, O. (2015). Unconventional gas development in the USA: Exploring the risk perception issues. *Risk Analysis, 35*(10), 1770–1788. https://doi.org/10.1111/risa.12512

Gramling, R., & Freudenburg, W. R. (1992). Opportunity-threat, development, and adaptation: Toward a comprehensive framework for social impact Assessment1. *Rural Sociology, 57*(2), 216–234. https://doi.org/10.1111/j.1549-0831.1992.tb00464.x

Hager, C., & Haddad, M. A. (2015). *NIMBY is beautiful: Cases of local activism and environmental innovation around the world*. Berghahn Books.

Haggerty, J. H., Kroepsch, A. C., Walsh, K. B., Smith, K. K., & Bowen, D. W. (2018). *Geographies of impact and the impacts of geography: Unconventional oil and gas in the American west*. The Extractive Industries and Society.

Haggett, C. (2011). Understanding public responses to offshore wind power. *Energy Policy, 39*(2), 503–510. https://doi.org/10.1016/j.enpol.2010.10.014

Harlan, S. L., Pellow, D. N., Roberts, J. T., Bell, S. E., Holt, W. G., & Nagel, J. (2015). Climate justice and inequality. *Climate Change and Society: Sociological Perspectives*, 127–163.

Hazboun, S. O., & Boudet, H. S. (2020). Public preferences in a shifting energy future: Comparing public views of eight energy sources in north America's Pacific Northwest. *Energies, 13*(8), 1940. https://www.mdpi.com/1996-1073/13/8/1940.

Hazboun, S. O., & Boudet, H. S. (2021). Natural gas—friend or foe of the environment? Evaluating the framing contest over natural gas through a public opinion survey in the Pacific Northwest. *Environmental Sociology, 1–14.*

Ho, S. S., Leong, A. D., Looi, J., Chen, L., Pang, N., & Tandoc, E. (2018). *Science literacy or value predisposition? A meta-analysis of factors predicting public perceptions of benefits, risks, and acceptance of nuclear energy*. Environmental Communication. https://doi.org/10.1080/17524032.2017.1394891

Inhaber, H. (1998). *Slaying the NIMBY dragon*. Routledge.

International Energy Agency. (2021a). *A rebound in global coal demand in 2021 is set to be short-lived, but no immediate decline in sight*. https://www.iea.org/news/a-rebound-in-global-coal-demand-in-2021-is-set-to-be-short-lived-but-no-immediate-decline-in-sight.

International Energy Agency. (2021b). *Global Energy Review 2021: Assessing the effects of economic recoveries on global energy demand and CO_2 emissions in 2021*. https://iea.blob.core.windows.net/assets/d0031107-401d-4a2f-a48b-9eed19457335/GlobalEnergyReview2021.pdf.

International Energy Agency. (2021c). *Oil 2021: Analysis and forecast to 2026*. International Energy Agency. https://www.iea.org/reports/oil-2021.

International Energy Agency's. (2021d). *Statement on recent developments in natural gas and electricity markets*. https://www.iea.org/news/statement-on-recent-developments-in-natural-gas-and-electricity-markets, 12 October 2021. (Accessed 12 October 2021).

Jacquet, J. B. (2012). Landowner attitudes toward natural gas and wind farm development in northern Pennsylvania. *Energy Policy, 50*, 677–688. https://doi.org/10.1016/j.enpol.2012.08.011

Jacquet, J. B., Junod, A. N., Bugden, D., Wildermuth, G., Fergen, J. T., Jalbert, K., Rahm, B., Hagley, P., Brasier, K. J., Schafft, K., Glenna, L., Kelsey, T., Fershee, J., Kay, D. L., Stedman, R. C., & Ladlee, J. (2018). A decade of Marcellus Shale: Impacts to people, policy, and culture from 2008 to 2018 in the Greater Mid-Atlantic region of the United States. *The Extractive Industries and Society, 5*(4), 596–609. https://doi.org/10.1016/j.exis.2018.06.006

Jacquet, J. B., & Kay, D. (2014). The unconventional boomtown: Updating the impact model to fit new spatial and temporal scales. *Journal of Rural and Community Development, 9*(1), 1–23.

Jaspal, R., Nerlich, B., & Lemańcyzk, S. (2014). Fracking in the Polish press: Geopolitics and national identity. *Energy Policy, 74*, 253–261. https://doi.org/10.1016/j.enpol.2014.09.007

Jenkins, K., McCauley, D., Heffron, R., Stephan, H., & Rehner, R. (2016). Energy justice: A conceptual review. *Energy Research & Social Science, 11*, 174–182. https://doi.org/doi.org/10.1016/j.erss.2015.10.004.

Jerolmack, C., & Walker, E. T. (2018). Please in my backyard: Quiet mobilization in support of fracking in an appalachian community. *American Journal of Sociology, 124*(2), 479–516. https://doi.org/10.1086/698215

Junod, A. N., & Jacquet, J. B. (2019). Shale gas in coal country: Testing the Goldilocks Zone of energy impacts in the western Appalachian range. *Energy Research & Social Science, 55*, 155–167. https://doi.org/10.1016/j.erss.2019.04.017

Krannich, R. S., & Greider, T. (1984). Personal well-being in rapid growth and stable communities: Multiple indicators and contrasting results. *Rural Sociology, 49*(4), 541. https://www.proquest.com/scholarly-journals/personal-well-being-rapid-growth-stable/docview/1290830286/se-2?accountid=14026.

Krannich, R. S., Little, R. L., Mushkatel, A., Pijawka, K. D., & Jones, P. (1991). *Southern Nevada residents' views about the Yucca mountain high-level nuclear waste repository and related issues: A comparative analysis of urban and rural survey data (p. Medium: ED; size: 137 p.)*. Carson City, NV (United States): Nevada Nuclear Waste Project Office. https://www.osti.gov/servlets/purl/140781.

Kunreuther, H. C., Desvousges, W. H., & Slovic, P. (1988). Nevada's predicament public perceptions of risk from the proposed nuclear waste repository. *Environment: Science and Policy for Sustainable Development, 30*(8), 16–33. https://doi.org/10.1080/00139157.1988.9932541

Kunreuther, H. C., Easterling, D., Desvousges, W., & Slovic, P. (1990). Public attitudes toward siting a high-level nuclear waste repository in Nevada. *Risk Analysis, 10*(4), 469–484. https://doi.org/10.1111/j.1539-6924.1990.tb00533.x

Kunreuther, H. C., Kleindorfer, P., Knez, P. J., & Yaksick, R. (1987). A compensation mechanism for siting noxious facilities: Theory and experimental design. *Journal of Environmental Economics and Management, 14*(4), 371–383.

Kunreuther, H. C., & Lathrop, J. W. (1981). Siting hazardous facilities: Lessons from LNG. *Risk Analysis, 1*(4), 289–302. https://doi.org/10.1111/j.1539-6924.1981.tb01429.x

Kunreuther, H. C., & Linnerooth, J. (1984). Low probability accidents. *Risk Analysis, 4*(2), 143–152. https://doi.org/10.1111/j.1539-6924.1984.tb00943.x

Kunreuther, H. C., Linnerooth, J., Lathrop, J., Atz, H., Macgill, S., Mandl, C., Schwarz, M., & Thompson, M. (1983). *Risk analysis and decision processes: The siting of liquefied energy gas facilities in four countries*. Springer Science & Business Media.

Lachapelle, E., Kiss, S., & Montpetit, É. (2018). Public perceptions of hydraulic fracturing (Fracking) in Canada: Economic nationalism, issue familiarity, and cultural bias. *The Extractive Industries and Society, 5*(4), 634–647. https://doi.org/10.1016/j.exis.2018.07.003

Ladd, A. E. (2014). Environmental disputes and opportunity-threat impacts surrounding natural gas fracking in Louisiana. *Social Currents, 1*, 293–311. https://doi.org/10.1177/2329496514540132

Lerner, S. (2012). *Sacrifice zones: The front lines of toxic chemical exposure in the United States*. Mit Press.

Lesbirel, S. H. (1998). *NIMBY politics in Japan: Energy siting and the management of environmental conflict*. Cornell University Press.

Maldonado, J. K. (2018). *Seeking justice in an energy sacrifice zone: Standing on vanishing land in coastal Louisiana*. Routledge.

Mayer, A. (2016). Risk and benefits in a fracking boom: Evidence from Colorado. *Extractive Industries and Society, 3*(3), 744–753. https://doi.org/https://doi.org/10.1016/j.exis.2016.04.006.

McAdam, D., & Boudet, H. S. (2012). Putting social movements in their place: Explaining opposition to energy projects in the United States, 2000-2005. In *Cambridge studies in contentious politics*. Cambridge University Press. http://www.cambridge.org/gb/knowledge/isbn/item6636951/?site_locale=en_GB.

McAdam, D., Boudet, H. S., Davis, J., Orr, R. J., Richard Scott, W., & Levitt, R. E. (2010). "Site fights": Explaining opposition to pipeline projects in the developing World1. In *Sociological forum* (Vol. 25, pp. 401–427). Wiley Online Library.

McCright, A. M., & Dunlap, R. E. (2013). Bringing ideology in: The conservative white male effect on worry about environmental problems in the USA. *Journal of Risk Research, 16*(2), 211–226.

Mohai, P., Pellow, D., & Roberts, J. T. (2009). Environmental justice. *Annual Review of Environment and Resources, 34*, 405–430.

Molotch, H. (1970). Oil in santa barbara and power in America*. *Sociological Inquiry, 40*(1), 131–144. https://doi.org/10.1111/j.1475-682X.1970.tb00990.x

Molotch, H., & Lester, M. (1975). Accidental news: The great oil spill as local occurrence and national event. *American Journal of Sociology, 81*(2), 235–260. https://doi.org/10.1086/226073

O'Hare, M., & Sanderson, D. (1993). Facility siting and compensation: Lessons from the Massachusetts experience. *Journal of Policy Analysis and Management, 12*(2), 364–376. https://doi.org/10.2307/3325241

Pierce, J. J., Boudet, H., Zanocco, C., & Hillyard, M. (2018). Analyzing the factors that influence U.S. public support for exporting natural gas. *Energy Policy, 120*, 666–674. https://doi.org/10.1016/j.enpol.2018.05.066

Portney, K. E. (1984). Allaying the NIMBY syndrome: The potential for compensation in hazardous waste treatment facility siting. *Hazardous Waste, 1*(3), 411–421.

Rabe, B. G. (1994). *Beyond NIMBY: Hazardous waste siting in Canada and the United States*. Brookings Institution Press.

Rand, J., & Hoen, B. (2017). Thirty years of North American wind energy acceptance research: What have we learned? *Energy Research & Social Science, 29*, 135–148. https://doi.org/10.1016/j.erss.2017.05.019

Riley, E. D., Michael, E. K., & Eugene, A. R. (1993). *Public reactions to nuclear waste: Citizens' views of repository siting*. Duke University Press. doi:10.1515/9780822397731.

Roberts, J. T., & Toffolon-Weiss, M. M. (2001). *Chronicles from the environmental justice frontline*. Cambridge University Press.

Rosa, E. A., Machlis, G. E., & Keating, K. M. (1988). Energy and society. *Annual Review of Sociology, 14*, 149–172. https://doi.org/10.1146/annurev.so.14.080188.001053

Schively, C. (2007). Understanding the NIMBY and LULU phenomena: Reassessing our knowledge base and informing future research. *Journal of Planning Literature, 21*(3), 255–266. https://doi.org/10.1177/0885412206295845

Schlosberg, D. (2009). *Defining environmental justice: Theories, movements, and nature*. Oxford University Press.

Sherman, D. J. (2011). *Not here, not there, not anywhere: Politics, social movements, and the disposal of low-level radioactive waste*. Routledge.

Slovic, P. (1987). Perception of risk. *Science, 236*(4799), 280–285. https://doi.org/10.1126/science.3563507

Slovic, P., Layman, M., & Flynn, J. H. (1991). Risk perception, trust, and nuclear waste: Lessons from Yucca mountain. *Environment: Science and Policy for Sustainable Development, 33*(3), 6–30. https://doi.org/10.1080/00139157.1991.9931375

Smith, E. R. A. N., & Marquez, M. (2000). The other side of the NIMBY syndrome. *Society & Natural Resources, 13*(3), 273–280. https://doi.org/10.1080/089419200279108

Taylor, D. (2014). *Toxic communities*. New York University Press.

Thomas, M., Pidgeon, N., Evensen, D., Partridge, T., Hasell, A., Enders, C., Herr Harthorn, B., & Bradshaw, M. (2017). Public perceptions of hydraulic fracturing for shale gas and oil in the United States and Canada. *Wiley Interdisciplinary Reviews: Climate Change, 8*(3), e450. https://doi.org/10.1002/wcc.450

Tran, T., Taylor, C. L., Boudet, H. S., Baker, K., & Peterson, H. L. (2019). Using concepts from the study of social movements to understand community response to liquefied natural gas development in clatsop county, Oregon. *Case Studies in the Environment, 3*(1). https://doi.org/10.1525/cse.2018.001800

U. S. Energy Information Administration. (2021). The United States was a net total energy exporter in 2019 and 2020. In *U. S. energy facts explained* (Vol. 2021). https://www.eia.gov/energyexplained/us-energy-facts/imports-and-exports.php.

Vasi, I. B., Walker, E. T., Johnson, J. S., & Tan, H. F. (2015). "No fracking way!" documentary film, discursive opportunity, and local opposition against hydraulic fracturing in the United States, 2010 to 2013. *American Sociological Review, 80*(5), 934–959. https://doi.org/10.1177/0003122415598534

Walsh, E. J. (1981). Resource mobilization and citizen protest in communities around three mile Island. *Social Problems, 29*(1), 1–21.

Wolsink, M. (2006). Invalid theory impedes our understanding: A critique on the persistence of the language of NIMBY. *Transactions of the Institute of British Geographers, 31*(1), 85–91.

Wolsink, M. (2007). Wind power implementation: The nature of public attitudes: Equity and fairness instead of 'backyard motives. *Renewable and Sustainable Energy Reviews, 11*(6), 1188–1207.

Wright, R. A., & Boudet, H. S. (2012). To act or not to act: Context, capability, and community response to environmental risk. *American Journal of Sociology, 118*(3), 728–777. https://doi.org/10.1086/667719

Wüstenhagen, R., Wolsink, M., & Bürer, M. J. (2007). Social acceptance of renewable energy innovation: An introduction to the concept. *Energy Policy, 35*(5), 2683–2691.

Zanocco, C., Boudet, H., Clarke, C. E., & Howe, P. D. (2019). Spatial discontinuities in support for hydraulic fracturing: Searching for a "goldilocks zone. *Society & Natural Resources, 32*(9), 1065–1072. https://doi.org/10.1080/08941920.2019.1616864

Zanocco, C., Boudet, H., Clarke, C. E., Stedman, R., & Evensen, D. (2020). NIMBY, YIMBY, or something else? Geographies of public perceptions of shale gas development in the marcellus shale. *Environmental Research Letters, 15*(7). https://doi.org/10.1088/1748-9326/ab7d01

https://www.iea.org/news/statement-on-recent-developments-in-natural-gas-and-electricity-markets, (2021). (Accessed 12 October 2021).

PART II

The new landscape of fossil fuel technology, supply, and policy

CHAPTER 2

The new global energy order: shifting players, policies, and power dynamics

Farid Guliyev
Department of Political Science and Philosophy, Khazar University, Baku, Azerbaijan

Introduction: the emergence of a new energy order

The crisis consists precisely in the fact that the old is dying and the new cannot be born; in this interregnum a great variety of morbid symptoms appear.

Antonio Gramsci in 1930 (Gramsci, 1971).

Alarming rates of anthropogenic global warming have made it imperative for governments across the globe to adopt climate mitigation policies to reduce greenhouse gas (GHG) emissions, including those from the fossil fuel industry—the biggest emitter of harmful CO_2 emissions and contributor to the earth's warming. This clean energy transition entails a profound structural shift from an energy system based on fossil fuels to one that relies primarily on (clean) renewable sources. Energy policy has become inextricably linked to climate policy, constituting the global climate–energy nexus. While the urgency of addressing the climate crisis is well recognized, climate change concerns have been addressed to a varying degree of efficacy at both national and global levels, including through the Paris Climate Agreement which aims to curb carbon emissions and limit global temperature "to well below 2°C" (United Nations Framework Convention on Climate Change (UNFCCC), 2015).

Progress is uneven though. For some regions, like the European Union, the low-carbon transition has become a policy priority. In January 2020, the European Parliament approved an ambitious "Green Deal" plan aimed at achieving climate neutrality by 2050 (European Commission, 2020). According to the World Economic Forum's Energy Transition Index—which measures both current energy system performance and the enabling environment for an energy transition—the Nordic countries plus Switzerland, Austria, the United Kingdom, France, and the Netherlands are leading the world in sustainable energy management and renewables

Public Responses to Fossil Fuel Export
ISBN 978-0-12-824046-5
https://doi.org/10.1016/B978-0-12-824046-5.00004-7

Table 2.1 Energy Transition Index 2020—Top 20 ranked by performance.

Country	ETI Ranking
Sweden	1
Switzerland	2
Finland	3
Denmark	4
Norway	5
Austria	6
United Kingdom	7
France	8
Netherlands	9
Iceland	10
Uruguay	11
Ireland	12
Singapore	13
Luxembourg	14
Lithuania	15
Latvia	16
New Zealand	17
Belgium	18
Portugal	19
Germany	20

(World Economic Forum (WEF), 2020) (see Table 2.1). Other countries—like China and India—are making steady progress, while still others—like the United States, Canada, and Brazil—are making little progress or even declining in their scores on the index.

While many countries have pledged to cut emissions, it is unlikely that the target of net-zero carbon dioxide (CO_2) emissions will be met without accelerating clean technological innovations (International Energy Agency (IEA), 2020). Most energy efficiency and clean energy technologies are concentrated in the electricity sector, where some jurisdictions have achieved impressive results. In Germany, for example, renewables today account for half of electricity consumption owing to the federal government's support for renewable adoption (Waldholtz, 2020). In Australia, the City of Sydney runs on 100% renewables (Nogrady, 2019). Yet, as impressive as these individual efforts may look, new technologies need to be developed to substantially cut emissions in other sectors such as aviation, heavy industries, and cargo shipping, which are responsible for a large chunk of emissions. The COVID-19 crisis and the downturn in energy markets may slow down the development, adoption, and deployment of low-carbon technologies (IEA, 2020), although the EU's 750 billion Euro recovery package prioritizes clean energy investment (European Parliament, 2020).

The scale and level of global dependence on fossil fuels increases the likelihood that the ongoing energy transition will be a "gradual, prolonged affair" (Smil, 2019). Though the pathways to clean energy will be winding and the transition will likely advance in an incremental, piecemeal fashion, even a modest reduction in fossil fuels in primary energy consumption is expected to have potentially far-reaching geostrategic implications. A high priority given to energy efficiency and climate investments in the EU is already reshaping global supply chains, reconfiguring former linkages and dependences between exporters and importers, bringing in new market players, and undercutting the power and resources of major oil producing states and fossil fuel corporations. Renewable energy heralds a more decentralized mode of energy generation and distribution by a more varied set of private businesses, making renewables more conducive to a competitive market structure. The clean energy transition, therefore, entails a shift from the currently dominant oligopolistic market structure for fossil fuels to a more decentralized and competitive one (Scholten, 2018).

As more and more advanced industrialized countries succeed with diversifying their energy mixes and reducing the intake of fossil fuels, they will become more energy self-sufficient, less dependent on traditional fossil fuel producers, and more resilient in the face of supply disruptions in the long run. As Overland (2019) states in the case of EU's energy-climate policy, if the EU succeeds in improving energy efficiency and manages to accelerate renewables deployment, they "will render fossil fuels a shrinking slice of a shrinking energy demand pie" (World Economic Forum (WEF), 2020).

However, a transition to renewables does not completely eliminate a set of issues traditionally associated with the "old" energy order, specifically availability, affordability, and security of energy supply; competition over access to critical pipeline and port infrastructure; and the use of energy resources as a foreign policy tool. Even though the share of renewables (wind, solar, biomass, and geothermal—excluding hydro) has increased from 4.5% in 2018 to 5% in 2019, fossil fuels still account for 84% of the world's energy mix (oil—33.1%, coal—27.0%, gas—24.2%) (BP, 2020a). Fossil fuel companies, both state-owned companies and international majors (plus smaller independent shale producers) and other companies along the supply chains, remain important players who will likely resist the climate change agenda for fear of losing market share to renewable companies and clean technology innovators. Traditional diesel-fueled car manufacturers (BMW, Volkswagen, Daimler) face competition from a new generation of electric car producers, such as Tesla (Fasse, 2019).

For the time being, these renewed geostrategic and social tensions manifest themselves in all sorts of crises, ruptures or disruptions, in what Gramsci (1971) [1930] called the "morbid symptoms" that are characteristic of any large-scale social change. Perhaps most prominent examples include the "oil price wars" over energy market share waged by traditional oil producers (Saudi Arabia and Russia) against independent shale companies in the United States and the controversy over the completion of the Nord Stream 2 pipeline to expand Russian gas deliveries to Western Europe (Becker, 2020).

Many issues remain uncertain as the process is complex, multifaceted, and still unfolding. However, it seems appropriate to use the concept of "transitional" or "shifting" to refer to the current energy order as one that combines the essential elements of both the "old" fossil fuel system and the "new" sociotechnological system of low-carbon sources and renewables. The concept of a "shifting order" allows one to avoid the extreme views and predictions of either energy transition skeptics like (Smil, 2019) who downplays the prospects of rapid decarbonization or cautious optimists like (Scholten et al., 2020) who build their assumptions on the thought experiment of a complete "renewables revolution." In some nations we could see a near complete transition rather soon, while in other nations status quo fossil fuel interests could retain structural dominance for much longer.

I argue that the energy order is stuck in-between the forces of fossil fuels and clean energy, and it is the interaction between these forces that will define global energy politics in decades to come. It is too early to pronounce the death of the fossil fuel industry, but it is also no longer possible to dismiss the advent of renewables. "Energy" is no longer the prerogative of fossil fuels, as new actors have emerged with stakes in clean energy policy seeking to limit the dominance of the conventional fossil fuel–based model. The current order is a mix of the two, and their nexus will determine the pathways of energy supply in the foreseeable future. In terms of policy implications, this means that the old geostrategic concerns with energy security and the resource curse will coexist with the so-called "geopolitics of renewables" (i.e., the impact of renewables on interstate relations) (Scholten et al., 2020)—and it is their *interaction* that adds the element of novelty.

To illustrate this point, consider the Middle East where most of the world's largest oil producers are concentrated. In the shifting energy order where fossil fuels coexist with renewables, large oil producers will eventually be pushed to gradually shift to wind, solar, and other renewables, requiring expensive technology and capital investments. Since the typical Gulf State economy is nondiversified with a large public sector and small (state-dependent and rent-seeking) private sector (Malik & Awadallah, 2013), Gulf governments will be pressed to channel large sums of oil export earnings to acquire renewables. As the price of oil is expected to decline, the Gulf States—where cost of oil is lowest—will have to pump more oil to increase the profits from the sale of oil for renewable investment (Nakhle, 2020). In short, the typical problems of petroleum dependence (lack of economic diversification, sunk costs associated with historic path-dependent choices) will interact with new challenges relating to the clean energy transition (technology and knowledge transfer, clean energy investment) in peculiar ways. In other regions with little or no oil production, the interplay of fossil fuels with clean energy will play out differently producing a different set of dynamics. Countries with ambitious climate mitigation policies in place are expected to incrementally phase out the use of carbon-intensive fossil fuel (such as coal) and shift to low-carbon alternatives. However, large sunk costs (plants and facilities) associated with the fossil fuel industry will inevitably slow down the adoption and switch to renewables (Fattouh et al., 2019).

Renewable technologies also require access to rare minerals, such as cobalt and lithium used in the production of solar photovoltaics and wind turbines. In other words, the energy transition will be mining intensive, replacing one set of environmental problems (carbon emissions) with another set of problems (how to ensure sustainable mining), fueling competition over access to such "critical" minerals and creating negative spillover effects, including environmental and sustainability concerns (Sovacool et al., 2020). Moreover, the high concentration of technological innovation in the Global North and high costs of renewable development make their deployment in the Global South unaffordable, even though solar radiation is more plentiful in the Global South.

Global energy system

COVID-19 energy impacts

During the spring of 2020, as the highly contagious novel coronavirus (COVID-19) spread across the globe, the world "stood still," and energy markets were in turmoil. Some expected that the global health pandemic would be a wake-up call for a more rapid shift to low-carbon energy and robust climate action. Others anticipated the economic shock would induce governments to drop a green recovery path and revert back to their traditional reliance on fossil fuels (Hanna et al., 2020). How has the COVID-19 pandemic impacted the energy sector?

Oil prices went through two rounds of extreme volatility. WTI crude price fell from $60/barrel in late 2019 to negative $37 for the May WTI futures contract and then stabilized in the $30—40 range in May—June 2020 (Jaffe, 2020).

The pandemic has caused a major economic disruption, considerable health and social impacts comparable in scale to the 1929 economic collapse (Great Depression), and the oil price shock in 1973 (McKee & Stuckler, 2020; OECD, 2020). The world GDP growth is expected to contract by almost 1%, in the worst-case scenario analysis (United Nations (UN), 2020).

The pandemic has had the "most severe" impact on the energy industry (Victor, 2020) and is considered unprecedented in history (IEA, 2020). Direct effects of the coronavirus crisis include the drop in demand and company financial pressures.

First, travel bans and lockdowns imposed worldwide to curb the spread of the coronavirus led to a freeze in economic activities, dampening the demand for fossil fuels, particularly oil. According to OPEC estimates, global oil demand could decline by about 9.8 million barrels a day (mmb/d) for the entire year, whereas total global oil demand is estimated to be at 90.0 mmb/d (OPEC, 2020). As a result, an oil producer is left to decide whether to keep producing at existing levels (accumulating extra oil in storage if such capacity is available) despite the sharp decline in demand or to cut production levels to avoid market saturation and a concomitant drop in oil prices (World Bank, 2020).

Responses to falling demand (and declining oil price) varied, but the so-called "oil price war" between Saudi Arabia and Russia has been at the center of global media coverage. In response to the COVID-19 crisis, Saudi Arabia-led OPEC called for a meeting with Russia and a host of smaller producers to negotiate oil production cuts, the traditional instrument used by the oil cartel in the past with varied success. Previously, Russia withdrew from its production quota commitments in the framework of OPEC+, sending oil prices in free fall. In the meeting held on March 6, OPEC and Russia failed to agree on production cuts. The next day, oil prices collapsed, hitting the lowest level since the financial crisis in 2008. Brent crude price declined from $55.48 per barrel in late February to an average of $33.73 per barrel in March 2020. Saudi Arabia responded by flooding the energy markets with cheaper oil, hoping that keeping oil prices at low levels long enough would induce Russia to return to a negotiating table. All oil-producing countries, whose economies were already struggling due to the pandemic, suffered considerably.

One month on, OPEC+ met again and on April 13 finally agreed on what was described as "historic" 9.7 mmb/d production cuts over 2 months (May–June 2020) (Cho & Murtaugh, 2020). Despite the large cuts, the energy markets saturation, low demand, and the continued lockdowns across major economies left oil prices seemingly unaffected. The most striking aspect of the OPEC+ deal was the role of American president Donald Trump, who acted as a broker pressuring Saudi Arabia and Russia to agree on new cuts. There are two explanations for this shift in the US position, which traditionally opposed high oil prices and never aligned with OPEC. First, the shale revolution in the United States turned the country into a large exporter of shale oil and gas. While generally low prices benefit US firms and consumers, they hurt its oil industry, especially more vulnerable shale producers. Second, the powerful fossil fuel lobby and well-documented role of interest groups in American policy. The US fossil fuel lobby is well funded and organized, although its preferences may not be in line with the preferences of most US voters.

Then US Secretary of Energy Dan Brouillette linked the OPEC+ production cut deal to the presumed benefits of the shale revolution that made the United States "more self-sufficient" (quoted in Sevastopulo & Brower, 2020). Before the 2020 OPEC+ oil cut deal, the United States was considering the protectionist option of blocking oil imports from Saudi Arabia and imposing tariffs on foreign oil imports following active lobbying efforts by US shale companies whose interests would suffer in the eventuality of a Saudi plan to dump oil markets with cheap oil (Brower & Meyer, 2020).

Secondly, the pandemic has had a negative impact on producers' financial balance sheets. In the face of falling demand, some companies, especially smaller ones, halted extraction, abandoned rigs, and stopped further exploration activities. As a result, oil companies face large income losses and low (or zero) returns to investors. In June 2020, BP announced it would reduce the value of its assets by $17.5 billion in expectation of long-term low crude prices (BP, 2020b).

Shale companies in the United States, which had already been in a precarious financial situation before the pandemic, were hit the worst and some are now on the brink of bankruptcy considering that shale is commercially unprofitable at prices below $50 a barrel. Total oil production fell to 10.5 mmb/d. Chesapeake Energy, a pioneer in shale, filed for bankruptcy (Brower & McCormick, 2020). Some industry experts noted that after the pandemic the shale revolution is over. The collapse in oil prices and storage problems led many shale producers to shut in most of their wells in May 2020. Tight oil output fell to 7.6 mmb/d in July, and it is unlikely to return to the pre-Covid 12 mmb/d level (Berman, 2020; Energy Information Administration (EIA), 2020).

Between old and new

The current period is unusual also because the COVID-19 pandemic coincides with the global energy transition, which when completed, promises to eliminate above-mentioned concerns with oil price volatility, geopolitical competition, and fossil fuel supply-chain disruptions. In a fully decarbonized world, these issues will lose relevance and become obsolete.

It is not the first time that human civilization is going through an energy transition. Previous global energy shifts (Smil, 2017) were shaped primarily by multilevel competition between "great powers" over the source of energy supply, rivalry between oil companies over a market share, and social conflict involving environmental groups. Such shifts can be conceived of as part of long-wave developments, such as the Kondratieff cycle-type of "technological regime shifts" or Schumpeterian waves of creative destruction (Podobnik, 1999).

For example, coal dominated during the Industrial Revolution in the late 18th to 19th century. Then, from World War I through the end of World War II, the world energy industry experienced a major shift toward petroleum. All major industrialized economies switched to petroleum by around the early 1970s. The share of petroleum rose from 5% of world energy supply in 1910 to around 50% in 1973 (Podobnik, 1999). As a result, fossil fuels became the primary source of energy that has driven industrial growth and underpins the modern democratic system (Mitchell, 2011) (See Figs. 2.1 and 2.2 below Ritchie & Roser, 2019).

At the turn of the 21st century, a commodity super cycle (2004–14) was inextricably linked to the increased flow of fossil fuel exports from the big oil producers (Saudi Arabia, Russia, Iran) to rising markets in China and other emerging economies. Their growing energy appetite sustained high oil prices for almost a decade (Hughes & Lipscy, 2013). That oil boom not only empowered traditional oil producers—a group of 13 oil producing OPEC nations plus Russia (and their mostly authoritarian rulers)—it was also largely congruent with the global energy order established in 1970s.

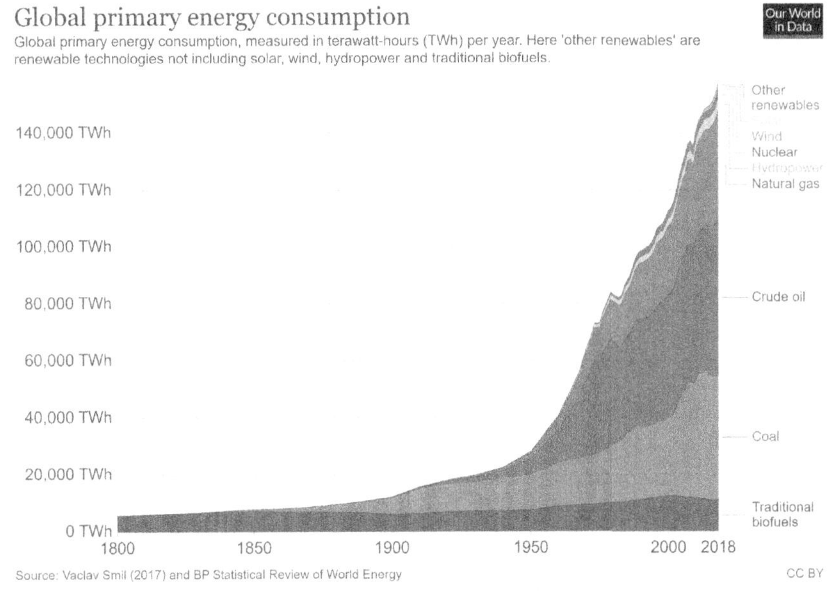

Figure 2.1 Global primary energy consumption. *(From Ritchie, H., & Roser, M. (2019).* Energy. *Published online at OurWorldInData.org. https://ourworldindata.org/energy.)*

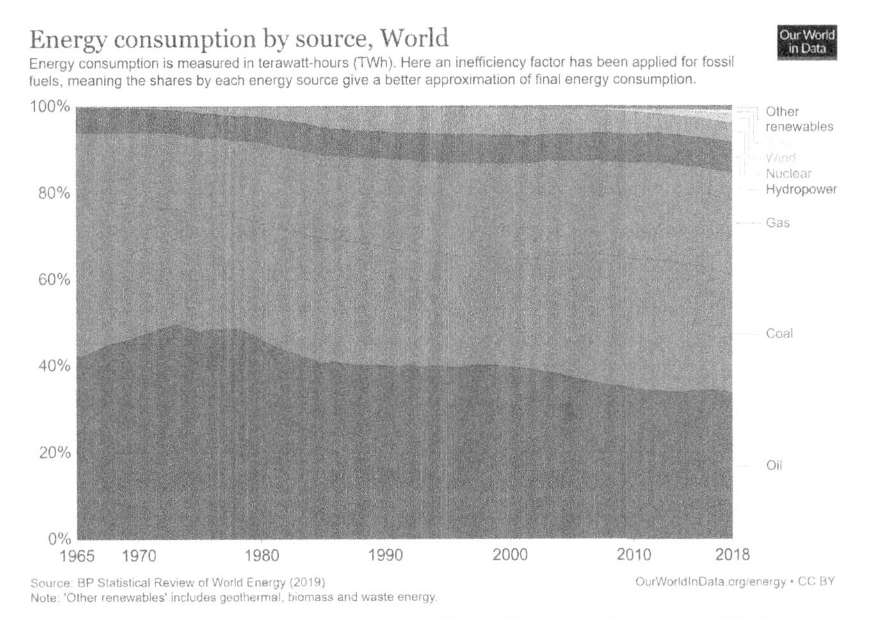

Figure 2.2 Energy consumption by source. *(From Ritchie, H., & Roser, M. (2019).* Energy. *Published online at OurWorldInData.org. https://ourworldindata.org/energy.)*

Before the oil crises that hit the world in the 1970s, global oil was dominated by several major international oil companies (IOCs) called the "Seven Sisters" or the Super Majors—Chevron, British Petroleum, ConocoPhillips, Royal Dutch Shell, Total, and ExxonMobil (Ross, 2013; Yergin, 2011). Throughout the 1970s, however, a dozen of major oil producing countries with large hydrocarbon reserves from Venezuela to Saudi Arabia and Iran all nationalized oil by taking control of their subsoil resources and creating state-owned national oil companies (NOCs). Oil nationalizations allowed governments in those countries to increase the share of resource rents accrued to the government. Having concentrated all oil wealth in the outsized public sector, some of these so-called rentier states like Saudi Arabia, the Gulf States, and Iran have had very little incentive to develop private productive activities outside the fossil fuel industry. This phenomenon has come to be known as the "paradox of plenty" (Karl, 1997).

In response to competition with NOCs, international majors (IOCs) went through a series of mergers and consolidation and became more globalized (Yergin, 2011). However, NOCs replaced IOCs in dominating global oil and gas reserves with 77% control of all oil and gas reserves (compared to the pre-1970s nationalizations when the Seven Sisters controlled 88% of global oil trade). NOCs have grown in size and influence over the years and some oil assets were renationalized in places like Russia (the Yukos affair) and Venezuela (where US oil giants ExxonMobil and ConocoPhillips were pushed out of the country in 2007 amidst Chavez's heightened intolerance of the United States) (Ellsworth, 2007; World Energy Council, 2016).

While oil export has always been a goal of development in the traditional oil producers, it has only recently been added to the resource development goals in the United States. In late 2008, the use of new technologies of horizontal drilling and hydraulic fracturing in the United States led to the surge in shale oil and gas (tight oil) development—the shale boom flourished until the end of 2014 (Frondel & Horvath, 2019). Bolstered by domestic shale production, the Trump administration promoted the vision of the United States as a major energy producer (Guliyev, 2020) that would reduce US dependence on imported oil and gas (especially on Middle Eastern sources) (Medlock et al., 2011).

Another cycle of volatility began in 2014 when, due to oversupply and declining demand in emerging markets, oil prices fell, sending a shockwave across oil-producing regions in the MENA, Russia and Eurasia, the Gulf of Guinea (in sub-Saharan Africa), Latin America (Venezuela, Bolivia), and other major oil-producing regions (shale producing states in the United States such as North Dakota, Texas). While oil prices were showing signs of recovery in 2018–19, a dramatic drop in global energy demand amidst the COVID-19 pandemic pushed prices to fall to record-low levels in April–May 2020.

There is a growing recognition that the global energy order is undergoing a major shift from the old system (which was in part "inherited" from the oil shocks of 1970s). However, precisely because the energy transition is unfolding, we are yet to fully understand and appreciate the full extent, nature, dynamics, and implications of this change for both producers and importers.

Clean energy transition

The concept "energy transition" refers to "the change in the composition (structure) of primary energy supply, the gradual shift from a specific pattern of energy provision to a new state of an energy system" (Smil, 2010). The current energy transition is one in which countries seek to shift energy supply away from fossil fuels (coal, oil, and natural gas) to renewable sources (wind, geothermal, solar, and biofuels). Goldthau et al. (2020) define the ongoing clean energy transition as "full decarbonization of the world energy system until 2050".

We are still in the early stages of this decarbonization process, which requires a complex, multifaceted sociotechnical shift that is expected to take decades to materialize (Smil, 2019). While significant strides have been made, especially in electricity generation, the speed and scale of the advancement of renewables is not currently sufficient to mitigate climate change targets. In world energy statistics, conventional fossil fuels are still the dominant source of energy, accounting for 81% of global primary energy consumption (oil—32%, coal—26%, gas—23%), while the remaining 19% are supplied by hydroelectric, biofuels, wind and solar power, and other renewables (Enerdata, 2019).

Natural gas was seen by western policymakers as an alternative to coal and as a bridge to renewable energy. The growth in gas demand made the EU increasingly more dependent on Russian gas giant Gazprom (at around 30 of net imports). Russia is believed to have used its gas as a geopolitical weapon. While some EU member states like Germany and Sweden, which are more advanced in renewables, are thought to be less vulnerable to Russia's gas leverage, the EU's Eastern European and Baltic members lag behind on decarbonization and are thus more dependent on Russian gas supply (Rodríguez-Fernández et al., 2020). In the United States, during its recent shale boom, oil and gas production has surged dramatically. Annual shale gas production grew from estimated 1 trillion cubic feet (Tcf) in 2006 to 9.7 Tcf in 2012. The increase in shale gas contributed to lower price for consumers, while the abundance of shale natural gas replaced coal for US electricity generation which could have lowered CO_2 emissions (Mason et al., 2015). Despite some of these advantages, however, natural gas exhibits all the negative patterns that climate mitigation policy seeks to tackle and represents a continuation of the old, fossil fuel–dominated regime.

It has become clear that the clean energy transition is uneven and does not encompass all countries to a similar extent (Fattouh et al., 2019). The process is best viewed as a "geographical process" with different scale and spatial patterns of distribution (Bridge et al., 2013).

While the European Union leads the world in the deployment of renewables (with the current target of 27% renewables by 2030 and net-zero by 2050 under European Green Deal), traditional oil producers in the Middle East, Russia, and Eurasia as well as many parts of the Global South (Venezuela, Nigeria) lag behind and continue to rely on fossil fuels almost exclusively. China is the world's biggest consumer of energy and its largest carbon-emitting powerhouse. Most of its energy consumption comes from fossil fuels (58% coal, 20% oil, and 8% natural gas) and renewables accounted for about 13% of total energy consumption in 2019 (EIA, 2020). In 2016, the Chinese government set out a plan to increase the share of renewables (solar and wind) to 20% by 2030 (O'Meara, 2020).

Countries such as Germany have achieved a great deal of progress. Through the decisive climate action called the *Energiewende*, Germany is on the path to phase out nuclear and a GHG emissions reduction by 80% by 2050. By encouraging renewables in power generation, renewables today account for half of Germany's electricity use. For comparison, in the United States 17% of electricity comes from renewables (Gielen et al., 2019). The introduction and advancement of a clean energy agenda is also unevenly spread within countries: for example, in the United States while some states such as California championed renewables, other states (e.g., Texas, North Dakota) have granted a green light to shale producers, and the federal government retreated on international climate change commitments—in 2018, President Trump announced the US exit from the Paris Agreement. Biden recently announced plans to rejoin.

Biden's cabinet appointments as well as the establishment of new positions like a special presidential envoy for climate change and the White

House national climate advisor signal the shift in US leadership's approach to tackling climate change (Ladislaw, 2021; White House, 2021). While this creates a window of opportunity for climate action, the new administration's ambitious energy and climate agenda plans will likely be disrupted in Congress and meet opposition by the powerful fossil lobbies (Pflüger, 2021). Changes, therefore, may be incremental (when a climate crisis requires more radical actions).

Oil-producing countries are generally less willing to implement clean energy transition policies—not least due to large sunk costs, legacy infrastructure, and, more importantly, the power of vested interests to block climate policy. Russia, the world's largest producer of oil and gas is a good case in point. In similarity to conservative-minded groups in the United States (including President Trump and his supporters), Russian political leaders cast doubt on global warming or deny climate change. Despite the potential to increase the share of renewables from about 5% to 11.3% of total energy consumption by 2030, Russia has a very low share of renewables. The vast amount of reserves of oil, gas, and coal disincentivizes Russia from enforcing climate mitigation efforts and the use of clean energy (Mitrova & Melnikov, 2019).

The uneven nature of the current clean energy transition entails the following:

(1) the regional, and resource-constrained, pathways characterized by the cohabitation of fossil fuels and renewables in coming decades;

(2) developments in the energy sector will reflect the inherent contradiction between the two opposing forces: on the one hand, vested interests around the fossil fuel supply chains seeking to sabotage the shift to renewables and adapt to the sociotechnological disruption it is causing; on the other hand, the prorenewables forces are pushing for climate policies and abandonment of fossil fuels ("keep it in the ground");

(3) competition between the fossil fuel industry and its beneficiaries who seek to protect their stakes and privileges and clean energy producers and advocates pushing for the reduction (or elimination) of carbon footprint is likely dominate domestic political discourse in many Western countries especially in Europe.

The outcome remains uncertain. While the fossil fuel industry is likely to persist and retrench due to the path-dependent forces of carbon lock-in (Unruh, 2002), inertia and powerful vested interests, the shift to clean energy requires decisive action and resources, which many governments lack. External shocks such as the COVID-19 pandemic is a disruption that presents a potential window of opportunity for pushing for a fundamental change, as IEA director (Birol, 2020) pointed out. Yet, it is unclear what longer-term impacts the global pandemic will have on various actors' plans to switch to clean energy sources (Kuzemko et al., 2020).

As the process is still unfolding and transitions are uneven and many features remaining unknown, it is premature to talk about renewables as a new "geopolitical reality" (Scholten, 2018). What we are likely to see in the short to medium term is a transitional stage, in which the old world dominated by fossil fuel producers coexists with an emerging new realm of low-carbon sources and renewables. The old geostrategic concerns with the volatility of oil prices, import dependency, the security of supply will remain, but will be complemented by new concerns such as governments competing over so-called critical minerals, innovative technologies—in which richer advanced industrialized economies have a clear advantage over the Global South where countries lack funds and R&D (research and development) capacities and are likely to rely on conventional carbon-based sources and see slow transitions to renewables for the foreseeable future.

The current energy transition is historically unprecedented in the sense that if previous energy transition were driven by the voracious consumption of ever-expanding needs of the industry and humans, the new energy transition entails the deliberate attempts by various actors to limit the use of fossil fuels to mitigate human-caused climate change. Environmental advocates promote the use of clean technologies and a greater use of low-carbon sources such as solar, wind, and biofuels (for useful discussion, see Cherp et al., 2011; Kern & Rogge, 2016).

Winners and losers of the energy transition

As with any social disruptive innovation, the shift to low-carbon energy produces both winners and losers. Losers such as large international fossil fuel (extractive) companies and producing states have vested interests in the preservation of the carbon-based energy system. They will inevitably resist change or sabotage the transition to low-carbon sources that threatens to strand carbon wealth and associated capital investments in extraction, production, transportation, and transmission of oil, gas, and coal. Loss of carbon wealth is estimated in trillions of US dollars (Cust & Manley, 2018).

Prospective winners include renewables companies, clean tech, and electric battery and all-electric car manufacturers (mostly concentrated in the Silicon Valley), various transnational advocacy groups and ultimately all humanity. Considering that governments and extractive multinationals are well-organized, they have an advantage over less organized and more diffusely structured societies. As in the classic political economy of reform model, "losers [pro-status quo distributional coalitions] are well organized, while prospective winners face daunting collective action problems and are not" (Haggard & Kaufman, 1995).

This represents a collective action problem, i.e., although all societies will benefit from renewables and greener environment which represents a kind of public good, no one individually will seek to push for a change. This makes it important to put in place an international framework of norms and rules that would oversee a gradual phase out of fossil fuels and reduction in polluting emissions. While a global regime complex of energy—climate nexus exists, its effectiveness has been questioned (Keohane & Victor, 2011).

Stokes (2020) argues that fossil fuel companies and dominant power utilities were behind a campaign of denial and worked hard to block the clean energy transition in the United States, leading to the retrenchment of fossil fuels. At the same time, clean energy stakeholders lack power to counterbalance the fossil fuel interests, who vigorously oppose climate constraints. Expenditures on lobbying by the fossil fuel and utilities companies were much larger than expenditures by environmental organizations and renewable energy companies (Brulle, 2018). Over the years, the United States has largely failed to adopt and enforce federal climate policy, and in the past years there were rollbacks on environmental regulations (Stokes, 2020). Without active US participation—especially after the Trump Administration's decision to withdraw from climate-change multilateral arrangements (Bordoff, 2017), global efforts to mitigate climate change would be partial and ineffectual. The new Biden administration has formally rejoined the Paris agreement and announced the target to achieve net-zero emissions by 2050 which would require bipartisan cooperation and adoption of a more comprehensive legal framework by Congress (Bordoff, 2021).

In sum, renewable policy implementation depends on the power of vested interests. Fossil fuel interests are better organized and enjoy strong leverage to block a prospective structural shift to renewables (Moe, 2016). Referring to Mancur Olson's (1982) group size and collective action problem, policy can be "captured" by organized interests that erect legal and regulatory barriers to shield themselves against competitors, leading to "institutional rigidity" which is inimical to structural change (Moe, 2016).

Using a comparison of the Netherlands and Belgium, Ahmadov and Van der Borg (2019) show that petroleum (through its rent-seeking incentives) has a negative effect on renewables deployment in the Netherlands but not in Belgium, which has very little fossil fuel resources. They find the crucial difference in the role played by rent-seeking vested interests: in the Netherlands, petroleum companies such as Royal Dutch Shell had strong influence on government policy through lobbying, while renewable companies lacked economic power to lobby for a more extensive renewable adoption.

Conclusion

Key takeaways from this chapter:

- The clean energy transition is happening, but with varied speed and geographical spread—large variation in energy mixes (and targets) between different energy regions and subregions (e.g., more advanced in the European Union but still limited in Russia, Eurasia, and the Middle East; China still largely relies on fossil fuels for primary energy consumption, but has emerged as a key producer of clean energy technologies).
- The traditional energy model in which a small number of large companies produce, export, and distribute fossil fuels remains the modal energy market structure in some parts of the world. However, this model is being challenged by a new, more decentralized mode of renewable generation characterized by a multiplicity of localized actors and competitive market structure.
- The fossil fuel producers—both NOCs (such as Saudi Aramco and Gazprom) and multinationals (such as Chevron, Shell, and BP)—thus may be losing market share to renewables (in some regions but not universally), but they are here to stay and fight back.
- In the United States, the shale revolution enabled a decade of "energy dominance" but is unlikely to reduce energy imports dependence without decisive measures to expand renewables (in which the United States is lagging behind the EU), and the shale boom seems to have reached its turning point considering its financially unsustainable foundations.
- Ongoing energy transitions will hurt the traditional fossil fuel–producing countries—OPEC+ participating states—who might have to slash production and abandon reserves and infrastructure but the amount of wealth to be thus "stranded" will depend on the speed and extension of decarbonization.

- Given the persistent significance of fossil fuels, the conventional concerns with security of supply are likely to continue and will be supplemented with new ones relating to critical minerals used in clean technology.
- The shifting energy order will produce political security considerations peculiar to the concentration of renewables (technology) in the advanced industrialized West and the concentration of fossil fuels in OPEC+ countries. They lie at the intersection between the forces of fossil fuels and low-carbon advocacy and highlight the duality and contradiction inherent in the nature of the ongoing transition stage.

References

Ahmadov, A. K., & van der Borg, C. (2019). Do natural resources impede renewable energy production in the EU? A mixed-methods analysis. *Energy Policy, 126*, 361–369. https://doi.org/10.1016/j.enpol.2018.11.044

Becker, A. (2020). *Nord Stream 2: Who needs the Russian gas pipeline after all?* Deutsche Welle. September 9. https://www.dw.com/en/nord-stream-2-who-needs-the-russian-gas-pipeline-after-all/a-54849447

Berman, A. (2020). *The party is over for U.S. shale and energy dominance*. Forbes. June 17. https://www.forbes.com/sites/arthurberman/2020/06/17/the-party-is-over-for-shale-and-us-energy-dominance/#2cc78fe7122c

Birol, F. (2020). *Put clean energy at the heart of stimulus plans to counter the coronavirus crisis*. International Energy Agency. https://www.iea.org/commentaries/put-clean-energy-at-the-heart-of-stimulus-plans-to-counter-the-coronavirus-crisis.

Bordoff, J. (2017). Withdrawing from the Paris climate agreement hurts the US. *Nature Energy, 2*(17145), 1–3. https://doi.org/10.1038/nenergy.2017.145

Bordoff, J. (2021). *On climate, declaring "America is back" doesn't make it so. Foreign Policy*. February 26 https://foreignpolicy.com/2021/02/26/biden-climate-paris-agreement-congress-emissions-reductions/.

BP. (2020a). Press Release. June 15. https://www.bp.com/en/global/corporate/news-and-insights/press-releases/bp-revises-long-term-price-assumptions.html.

BP. (2020b). *Statistical review of world energy*. https://www.bp.com/en/global/corporate/energy-economics/statistical-review-of-world-energy/renewable-energy.html.

Bridge, G., Bouzarovski, S., Bradshaw, M., & Eyre, N. (2013). Geographies of energy transition: Space, place and the low-carbon economy. *Energy Policy, 53*, 331–340. https://doi.org/10.1016/j.enpol.2012.10.066

Brower, D., & McCormick, M. (2020). *Shale pioneer Chesapeake energy filed for bankruptcy*. Financial Times. June 28. https://www.ft.com/content/31f35631-7b7d-4fa1-8eef-6255bf18cf88

Brower, D., & Meyer, G. (2020). *US shale producers launch anti-Saudi lobbying push*. Financial Times. April 3. https://www.ft.com/content/af0c7876-10fa-4320-9d3a-7d913573ff56

Brulle, R. J. (2018). The climate lobby: A sectoral analysis of lobbying spending on climate change in the USA, 2000 to 2016. *Climatic Change, 149*(3–4), 289–303. https://doi.org/10.1007/s10584-018-2241-z

Cherp, A., Jewell, J., & Goldthau, A. (2011). Governing global energy: Systems, transitions, complexity. *Global Policy, 2*(1), 75–88. https://doi.org/10.1111/j.1758-5899.2010.00059.x

Cho, S., & Murtaugh, D. (2020). *Will historic OPEC+ cuts work? Oil timespreads say maybe not.* Bloomberg. April 13. https://www.bloomberg.com/news/articles/2020-04-13/will-historic-opec-cuts-work-oil-timespreads-say-maybe-not

Cust, J., & Manley, D. (2018). The carbon wealth of Nations: From rents to risks. In G.-M. Lange, Q. Wodon, & K. Carey (Eds.), *The changing wealth of nations: Building a sustainable future* (pp. 97–113). World Bank. https://doi.org/10.1596/978-1-4648-1046-6_ch5

EIA. (2020). China, last updated September 30. https://www.eia.gov/international/analysis/country/CHN.

Ellsworth, B. (2007). *Chavez drives Exxon and ConocoPhillips from Venezuela.* Reuters. June 26. https://www.reuters.com/article/uk-venezuela-nationalization-oil/chavez-drives-exxon-and-conocophillips-from-venezuela-idUKN2637895020070626

Enerdata. (2019). *Energy statistical yearbook.* https://yearbook.enerdata.net/total-energy/world-consumption-statistics.html.

Energy Information Administration (EIA). (2020). *U.S. field production of crude oil.* https://www.eia.gov/dnav/pet/hist/LeafHandler.ashx?n=PET&s=MCRFPUS2&f=M.

European Commission. (2020). *A European Green Deal: Striving to be the first climate-neutral continent.* https://ec.europa.eu/info/strategy/priorities-2019-2024/european-green-deal_en.

European Parliament. (2020). *Covid-19: EU recovery plan should prioritize climate investment.* https://www.europarl.europa.eu/news/en/headlines/society/20200429STO78172/covid-19-eu-recovery-plan-should-prioritise-climate-investment.

Fasse, M. (2019). Automotive crisis: Germany's car industry faces a perfect storm. *Handelsblatt.* February 22. https://www.handelsblatt.com/today/companies/automotive-crisis-germanys-car-industry-faces-a-perfect-storm/24026414.html

Fattouh, B., Poudineh, R., & West, R. (2019). The rise of renewables and energy transition: What adaptation strategy exists for oil companies and oil-exporting countries? *Energy Transitions,* 45–58. https://doi.org/10.1007/s41825-019-00013-x

Frondel, M., & Horvath, M. (2019). The US fracking boom: Impact on oil prices. *Energy Journal, 40*(4), 191–206. https://doi.org/10.5547/01956574.40.4.mfro

Gielen, D., Boshell, F., Saygin, D., Bazilian, M., Wagner, N., & Gorini, R. (2019). The role of renewable energy in the global energy transformation. *Energy Strategy Reviews, 24,* 38–50. https://doi.org/10.1016/j.esr.2019.01.006

Goldthau, A., Eicke, L., & Weko, S. (2020). The global energy transition and the global south. In M. Hafner, & S. Tagliapietra (Eds.), *The geopolitics of the global energy transition* (pp. 319–339). Springer International Publishing. https://doi.org/10.1093/acref/9780191843730.001.0001/q-oro-ed5-00018416

Gramsci, A. (1971). *Selections from the prison notebooks.* New York: International Publishers.

Guliyev, F. (2020). Trump's "America first" energy policy, contingency and the reconfiguration of the global energy order. *Energy Policy, 140,* 111435. https://doi.org/10.1016/j.enpol.2020.111435

Haggard, S., & Kaufman, R. (1995). *The political economy of democratic transitions.* Princeton: Princeton University Press.

Hanna, R., Xu, Y., & Victor, D. (2020). After COVID-19, green investment must deliver jobs to get political traction. *Nature, 582,* 178–180.

Hughes, L., & Lipscy, P. (2013). The politics of energy. *Annual Review of Political Science, 16*(1), 449–469. https://doi.org/10.1146/annurev-polisci-072211-143240

IEA. (2020). *The global oil industry is experiencing a shock like no other in its history.* https://www.iea.org/articles/the-global-oil-industry-is-experiencing-shock-like-no-other-in-its-history.

International Energy Agency (IEA). (2020). *Clean energy innovation.* https://www.iea.org/reports/clean-energy-innovation.

Jaffe, A. M. (2020). Geopolitics and the oil price cycle — an introduction. *Economics of Energy and Environmental Policy, 9*(2), 1–9. https://doi.org/10.5547/2160-5890.9.2.AJAF

Karl, T. L. (1997). *The paradox of plenty. Oil booms and petro-states.* University of California Press.

Keohane, R. O., & Victor, D. G. (2011). The regime complex for climate change. *Perspectives on Politics, 9*(1), 7–23. https://doi.org/10.1017/S1537592710004068

Kern, F., & Rogge, K. S. (2016). The pace of governed energy transitions: Agency, international dynamics and the global Paris agreement accelerating decarbonisation processes? *Energy Research & Social Science, 22,* 13–17. https://doi.org/10.1016/j.erss.2016.08.016

Kuzemko, C., Bradshaw, M., Bridge, G., & Goldthau, A. (2020). Covid-19 and the politics of sustainable energy transitions. *Energy Research & Social Science,* 101685. https://doi.org/10.1016/j.erss.2020.101685

Ladislaw, S. (2021). *Congress needs comprehensive agenda to battle climate change.* The Hill. February 18. https://thehill.com/opinion/energy-environment/539405-congress-needs-comprehensive-agenda-to-battle-climate-change?rl=1

Malik, A., & Awadallah, B. (2013). The economics of the Arab spring. *World Development,* 296–313. https://doi.org/10.1016/j.worlddev.2012.12.015

Mason, C. F., Muehlenbachs, L. A., & Olmstead, S. M. (2015). The economics of shale gas development. *Annual Review of Resource Economics,* 7(1), 269–289. https://doi.org/10.1146/annurev-resource-100814-125023

McKee, M., & Stuckler, D. (2020). If the world fails to protect the economy, COVID-19 will damage health not just now but also in the future. *Nature Medicine, 26*(5), 640–642. https://doi.org/10.1038/s41591-020-0863-y

Medlock, K., III, Jaffe, A. M., & Hartley, P. (2011). *Shale gas and US national security.* James Baker Institute for Public Policy, Rice University. https://scholarship.rice.edu/bitstream/handle/1911/91356/EF-pub-DOEShaleGas-07192011.pdf.

Mitchell, T. (2011). *Carbon democracy.* London: Verso.

Mitrova, T., & Melnikov, Y. (2019). Energy transition in Russia. *Energy Transitions, 3,* 73–80. https://doi.org/10.1007/s41825-019-00016-8

Moe, E. (2016). *Renewable energy transformation or fossil fuel backlash: Vested interests in the political economy.* Houndmills: Palgrave Macmillan.

Nakhle, C. (2020). *Clean energy and fossil fuels in the Middle East: A virtuous cycle? NRGI.* June 29. https://resourcegovernance.org/blog/clean-energy-fossil-fuels-middle-east-virtuous-cycle

Nogrady, B. (2019). Australia's capital city switches to 100% renewable energy. *Nature.* https://doi.org/10.1038/d41586-019-02804-0. September 19.

OECD. (2020). Coronavirus: The world economy at risk. *OECD Interim Economic Assessment.* http://www.oecd.org/berlin/publikationen/Interim-Economic-Assessment-2-March-2020.pdf.

Olson, M. (1982). *The rise and decline of nations.* New Haven, CT: Yale University Press.

O'Meara, S. (2020). China's plan to cut coal and boost green growth. *Nature,* S1–S3. https://doi.org/10.1038/d41586-020-02464-5

OPEC. (2020). *Monthly oil market report – November 2020.* https://momr.opec.org/pdf-download/res/pdf_delivery_momr.php?secToken2=accept.

Overland, I. (2019). EU climate and energy policy: New challenges for old energy suppliers. In J. Godzimirski (Ed.), *New political economy of energy in Europe* (pp. 73–102). Cham: Palgrave MacMillan.

Pflüger, F. (2021). *A paradigm shift under President Joe Biden: From "energy dominance" towards climate cooperation. Atlantic Council.* February 10. https://www.atlanticcouncil.org/blogs/energysource/a-paradigm-shift-under-president-joe-biden-from-energy-dominance-towards-climate-cooperation/

Podobnik, B. (1999). Toward a sustainable energy regime: A long-wave interpretation of global energy shifts. *Technological Forecasting and Social Change, 62*(3), 155–172. https://doi.org/10.1016/S0040-1625(99)00042-6

Ritchie, H., & Roser, M. (2019). *Energy*. Published online at OurWorldInData.org. https://ourworldindata.org/energy.

Rodríguez-Fernández, L., Fernández Carvajal, A. B., & Ruiz-Gómez, L. M. (2020). Evolution of European Union's energy security in gas supply during Russia–Ukraine gas crises (2006–2009). *Energy Strategy Reviews*, 100518. https://doi.org/10.1016/j.esr.2020.100518

Ross, M. L. (2013). *The oil curse*. Princeton University Press.

Scholten, D. (2018). *The geopolitics of renewables*. Cham: Springer.

Scholten, D., Bazilian, M., Overland, I., & Westphal, K. (2020). The geopolitics of renewables: New board, new game. *Energy Policy, 138*, 111059. https://doi.org/10.1016/j.enpol.2019.111059

Sevastopulo, D., & Brower, D. (2020). *US confident of 'fundamental shift' in oil politics*. Financial Times. April 14. https://www.ft.com/content/a0ed6215-7c65-4562-b211-1c1e9c414b6d?segmentId=98583035-ac35-a0ba-ed44-378e53f8caec

Smil, V. (2010). *Energy transitions: History, requirement, prospects*. Santa Barbara: Praeger.

Smil, V. (2017). *Energy transitions: Global and national perspectives*. Praeger.

Smil, V. (2019). What we need to know about the pace of decarbonization. *Substantia, 3*(2), 69–73. https://doi.org/10.13128/Substantia-700

Sovacool, B. K., Ali, S. H., Bazilian, M., Radley, B., Nemery, B., Okatz, J., & Mulvaney, D. (2020). Sustainable minerals and metals for a low-carbon future. *Science, 367*(6473), 30. https://doi.org/10.1126/science.aaz6003

Stokes, L. (2020). *Short circuiting policy: Interest groups and the battle over clean energy and climate policy in the American States*. Oxford University Press.

United Nations Framework Convention on Climate Change (UNFCCC). (2015). *Paris Agreement*. https://unfccc.int/process-and-meetings/the-paris-agreement/the-paris-agreement.

United Nations (UN). (2020). *World economic situation and prospects*. Briefing No. 136. https://www.un.org/development/desa/dpad/wp-content/uploads/sites/45/publication/Monthly_Briefing_136.pdf

Unruh, G. (2002). Escaping carbon lock-in. *Energy Policy, 30*(4), 317–325. https://doi.org/10.1016/s0301-4215(01)00098-2

Victor, D. (2020). *Forecasting energy futures amid the coronavirus outbreak*. Brookings. April 3. https://www.brookings.edu/blog/order-from-chaos/2020/04/03/forecasting-energy-futures-amid-the-coronavirus-outbreak/

Waldholtz, R. (2020). *Germany marks first ever quarter with more than 50 pct renewable electricity*. Clean Energy Wire. April 1 https://www.cleanenergywire.org/news/germany-marks-first-ever-quarter-more-50-pct-renewable-electricity.

White House. (2021). *Executive order on tackling the climate crisis at home and abroad*. https://www.whitehouse.gov/briefing-room/presidential-actions/2021/01/27/executive-order-on-tackling-the-climate-crisis-at-home-and-abroad/ (Original work published 2021).

World Bank. (2020). *Commodity prices tumbled further in March*. https://blogs.worldbank.org/opendata/commodity-prices-tumbled-further-march-pink-sheet.

World Economic Forum (WEF). (2020). *Energy transition index 2020*. https://www.weforum.org/reports/fostering-effective-energy-transition-2020.

World Energy Council. (2016). *World energy resources*. https://www.worldenergy.org/assets/images/imported/2016/10/World-Energy-Resources-Full-report-2016.10.03.pdf.

Yergin, D. (2011). *The prize: The epic quest for oil*. New York: Simon and Schuster.

CHAPTER 3

Fossil fuel export as a climate policy problem

Georgia Piggot[1,2] and Peter Erickson[2]
[1]School of Environment, University of Auckland, Auckland, New Zealand; [2]Stockholm Environment Institute US Center, Seattle, WA, United States

Burning fossil fuels is a key driver of the climate crisis. Emissions from coal, oil, and gas make up more than three-quarters of the anthropogenic greenhouse gas emissions that are causing global warming (IEA, 2019; SEI et al., 2020). Yet, despite the prominent role that fossil fuels play in climate change, the production and export of fossil fuels are rarely considered by governments as part of the remit of climate policy (Lazarus & van Asselt, 2018). Indeed, a paradox exists where some major coal, oil, and gas exporting countries have claimed climate leadership based on their efforts to reduce fossil fuel use domestically, even while the fuels they export generate damaging environmental impacts as they move throughout the supply chain worldwide (Bang & Lahn, 2020; Lee, 2018).

While many governments have long benefited from a global policy regime that allowed them to avoid responsibility for the climate impacts of their exports (Harrison, 2015), sentiment seems to be turning. Civil society has grown increasingly vocal over the past decade about the need to halt fossil fuel production and export to curb climate change (Blondeel et al., 2019; Cheon & Urpelainen, 2018; Piggot, 2018; Temper et al., 2020). Mobilization against exports, and the policies and financial institutions that support them, has shifted the conversation about climate change upstream to fossil fuel export facilities, which have become a site of climate contention (Harrison, 2020; Hazboun, 2019).

To understand how and why this contention matters, and why export is important in the context of climate governance, this chapter delves into the connection between fossil fuel production and climate policy. It explores how and why fossil fuel export got left off the global climate policy agenda, and discusses why it is important to bring the trade of fossil fuels into the climate policy process. This chapter provides context to understand the public responses highlighted throughout this volume, which have played a

Public Responses to Fossil Fuel Export
ISBN 978-0-12-824046-5
https://doi.org/10.1016/B978-0-12-824046-5.00013-8

critical role in shifting the policy landscape to bring the two previously distinct issues of climate and export together.

Why fossil fuel export is often ignored in climate policy

Fossil fuel export fell off the climate policy agenda as a consequence of policy design choices (Lazarus & van Asselt, 2018) and political maneuvering by fossil fuel producing nations aimed at avoiding discussion of their production activities (Depledge, 2008). Historical decisions about where in the supply chain to address the climate problem, how to account for emissions, and who ultimately bears responsibility for addressing those emissions combined to create a regime where fossil fuel production and export were treated as distinct from the climate policy challenge (Harrison, 2015).

Export is a marginalized issue in climate policy because emissions are typically addressed by the jurisdictions where fossil fuels are burned, rather than where they are produced or exported (with the exception of the portion of emissions that are released as a by-product of the fossil fuel extraction and production process). This focus leads policymakers to emphasize reducing emissions from the site of fossil fuel combustion (think power plants and tail pipes) rather than prevent the polluting fuels from being extracted and exported in the first place (Lazarus & van Asselt, 2018).

The way we account for emissions reinforces this focus on the point of combustion (Davis et al., 2011; Steininger et al., 2016). Governments worldwide have collectively agreed to use a "territorial" approach to emissions accounting, meaning that they count emissions released *within* their borders (IPCC, 2006). If fossil fuels produced in a territory are burned elsewhere, they fall off the accounting books of the country or region that produced them (Erickson & Lazarus, 2013). This practice gives little incentive to governments to address fossil fuel production and export, as they do not receive recognition for this as part of the accounting they complete in compliance with Paris Agreement commitments.

Recently, civil society organizations and climate scholars have suggested approaches to improve transparency about the alignment of their fossil fuel extraction plans with emissions reduction goals—either through a complementary set of extraction-based emissions accounts (Lee, 2018) or a tally of what fossil fuels are produced or expected to be produced (Byrnes, 2020). Such accounts would need to be in addition to the existing territorial accounting system, in order to prevent double-counting of emissions from fossil fuels when they are both extracted and burned. This type of accounting would provide a fuller picture of a country or region's contributions to global carbon dioxide emissions and provide a basis for discussions about where limiting production could help meet climate goals.

Finally, the global climate regime has historically prioritized nation-states as the foci for climate change negotiations and responses—though a trend toward greater inclusion of nonstate actors has occurred over time (Gupta, 2014). A consequence of this focus on national governments is that export facilities often fall through the cracks of climate change decision-making: local governments are typically responsible for permitting infrastructure (except where they cross jurisdictional boundaries or are significant enough to warrant national concern), but national governments are responsible for adhering to global climate commitments. For this reason, local governments who have attempted to limit fossil fuel exports—such as the City of Portland, Oregon, which introduced an ordinance banning new fossil fuel infrastructure in 2016, or Oakland, California, which banned the loading, storage, and handling of coal in the city's bulk facilities in 2016—have often faced legal challenges questioning their jurisdictional authority over climate, commerce, or energy decision-making (Perron, 2020).

In sum, the historical emphasis on fossil fuel combustion, territorial emissions, and national action in global climate policy has created blind spots around fossil fuel production and export. The design of the global climate regime has allowed countries to treat the export of fossil fuels as a benign climate act. This is beginning to shift—notably the UN Secretary General recently highlighted the need to focus on the transition away from fossil fuels when he stated "without a doubt that the production and use of coal, oil, and gas needs to decrease quickly if we are to achieve the goals of the Paris Agreement on climate change" (UNEP, 2020b). However, we still lack comprehensive policy frameworks in most major fossil fuel exporting nations and at the international level to address production and export.

The rationale for addressing fossil fuel export as a component of climate policymaking

A sole focus on fossil fuel combustion might be justified if the world had collectively cut demand for fossil fuels, hence making a complementary focus on supply redundant. But policymakers have not managed to successfully decarbonize the global economy and are presently on track to exceed warming by 3°C or more, far exceeding the targets of the Paris Agreement (UNEP, 2020a). The failure of the "idealized" global climate regime to effectively tackle emissions has many questioning whether new policy approaches may be helpful, including policies focused at the point of production or export of coal, oil, and gas (IEA, 2019; SEI et al., 2020).

The exploration for new and ambitious forms of climate policy led to proposals for "supply-side" climate policies that restrain the supply of fossil fuels (Gaulin & Le Billon, 2020; Lazarus & van Asselt, 2018). Indeed, some countries have begun to address fossil fuel production in earnest. The governments of Costa Rica, Belize, France, Denmark, New Zealand, Spain, and Ireland, for instance, have all banned exploration or extraction of oil in part or all of their territories (IEA, 2019; SEI et al., 2020). While we are yet to see major fossil fuel exporters commit to reducing production and export of fossil fuels to meet climate goals, some key producers are beginning to lay the groundwork for a transition away from coal, oil, and gas. For instance, the United States has recently paused oil and gas leasing on public lands and in offshore waters to allow the opportunity to review leasing arrangements in light of climate imperatives (The White House, 2021).

In the context of fossil fuel export, a range of supply-side policy options exist that could be used to limit export, such as revoking permits for export facilities (Rafaty et al., 2020), taxing fossil fuel exports (Antón, 2020), or simply banning fossil fuel exports altogether. Some subnational governments, for instance, have used zoning laws to prohibit permitting of fossil fuel export facilities within geographical boundaries (Perron, 2020). Governments can also change the calculus of decision-making so that climate change is given more weight, for instance, by requiring that upstream emissions are reported as part of the environmental review process for new export facility projects (Burger & Wentz, 2020). Or, they can limit their financial support for fossil fuel projects, by cutting subsidies to exporters or finance for new infrastructure projects (Moerenhout & Irschlinger, 2020). Finally, governments can facilitate the transition by providing support to help export-dependent communities diversify to alternative industries (Green & Gambhir, 2020).

Proponents of supply-side polices often argue that these policies should be introduced *alongside* efforts to cut emissions directly through measures such as carbon pricing or investments in energy efficiency. This dual focus would allow governments to "cut with both arms of the scissors," reducing both the supply of fossil fuels at the same time as demand (Green & Denniss, 2018). There are several reasons why this could help raise climate policy ambition and effectiveness more broadly.

An explicit focus on fossil fuel supply would help solve the current problem of *overproduction* of fossil fuels, which is hampering efforts to cut emissions. A recent report by UNEP examining plans from the world's largest fossil fuel producers and exporters found that governments are planning to produce 120% more fossil fuels by 2030 than would be consistent with limiting global warming to 1.5°C (IEA, 2019; SEI et al., 2020). Flooding the world with fossil fuels will make it harder to undertake the transition to renewable energy that is necessary to meet our climate goals.

Overinvestment in fossil fuels also creates the risk of "stranded assets"—fossil fuel infrastructure that becomes superfluous if countries switch to less polluting energy sources—which can cause huge liabilities for countries, companies, and communities that bet on a fossil-fueled future (Bos & Gupta, 2018; Leaton et al., 2013). Relatedly, by failing to tackle supply, we run the risk of exacerbating the problem of "carbon lock-in"—whereby our infrastructure, economy, politics, and culture become so enmeshed in a fossil-fueled way-of-life that it becomes challenging to envision and create alternatives (Erickson et al., 2015; Seto et al., 2016; Unruh, 2000). When communities decide to build out fossil fuel export terminals, for instance, they are making a long-term commitment to tie their economy to the fossil fuel industry. This "lock-in" can form a formidable barrier to building a low carbon future, as the relics of past decisions—such as the need to ensure a return-on-investment in infrastructure, the political power gained by incumbent industries, or the presence of a workforce trained for fossil fuel production—can inhibit the ability to shift to other, less-polluting industries. Preventing further build out of fossil fuel export facilities can limit exposure to this "lock-in," making the enactment of climate responses easier in the future.

A dedicated focus on fossil fuel supply in climate policy can also help ensure a more just and equitable transition to a climate-friendly future. If we only focus on building green industries as a climate response, we may forget about those regions, communities, and workers who currently depend on fossil fuel export for their survival. If instead, we also address high-carbon industries in climate policymaking, we better ensure that we have a well-managed transition to a low-carbon future that leaves no one behind. In the context of export, we may need to invest funds into decommissioning and cleaning up export sites, as well as retraining workers who currently work in fossil fuel exporting roles (Green & Gambhir, 2020). A "just transition" away from fossil fuels should also include support for communities that have been historical harmed by fossil fuel production and export, such as the fence line communities who have borne the brunt of pollution from neighboring fossil fuel facilities over the years (Healy et al., 2019). Low-income countries who rely heavily on fossil fuel export will also need assistance from the international community to help transition their economies (Armstrong, 2020; Muttitt & Kartha, 2020; Ross, 2019).

Addressing fossil fuel supply can help illuminate many of the blind spots that our current climate policy regime focused solely on emissions misses. Ultimately, nations of the world cannot, in aggregate, keep expanding production and export of fossil fuels, while also meeting their agreed climate goals. This requires grappling with the idea that fossil fuel exporting will be a less-prominent part of our future, and by planning for this fact now, we can ensure a more effective, timely, and just transition.

How public response has helped bridge the issues of climate change and fossil fuel export

Given the focus of this book, it seems relevant to conclude with a discussion of the role that the public has played in bringing the topic of fossil fuel production and export into the climate policy arena. Public responses—from protests to divestment campaigns, and lawsuits pushing for policy shifts—have been central in pushing the conversation about climate change upstream to include the production of the fuels that drive climate change (Cheon & Urpelainen, 2018; Piggot, 2018; Temper et al., 2020).

Indeed, all of the "first mover" cases, where government has acted to limit fossil fuel supply in the name of climate, have been underpinned by robust civil society mobilization calling for action to limit fossil fuel production and export (Carter & McKenzie, 2020; Tudela, 2020). In this book, for instance, Widener (2021) outlines the case of the 2018 New Zealand oil exploration ban, which was introduced in response to a civil society push to phase out fossil fuel production in the country. While the government stopped short of a full oil production ban (onshore production is still permitted), this mobilization catalyzed a new conversation that linked energy production and climate mitigation policy together for the first time in the country's history.

How is public mobilization on fossil fuels reshaping climate policy? First, and perhaps most critically, it is expanding discourse and norms about what climate leadership can look like. Public contention about fossil fuel export has made the issue more visible, making it difficult for governments to continue to claim their climate prowess while still expanding fossil fuel infrastructure. More broadly, the "keep it in the ground" movement has begun to catalyze a new "anti-fossil fuel norm," which makes proposing new export infrastructure (among all other types of fossil fuel infrastructure) a more challenging prospect (Blondeel et al., 2019; Green, 2018). As a result of these shifting norms, we now see leaders of major fossil fuel producing nations willing to talk about the need to transition away from oil, gas, and coal, such as when then candidate and now US President Joe Biden called for a "transition away from the oil industry" in his presidential debates in 2020—marking a notable shift in climate policy landscape (Friedman, 2020).

Public mobilization around fossil fuel production and export also appears to have made a tangible impact on climate change policymaking and emissions (albeit one that is difficult to quantify). Blockades and divestment campaigns have helped block or delay the construction of new fossil fuel infrastructure, which has slowed the supply of fossil fuels to market, ultimately putting a small dent in emissions (Erickson & Lazarus, 2014; Temper et al., 2020). But perhaps more importantly, campaigns against fossil fuel infrastructure have led to subtle shifts in policy processes that tilt the balance away from fossil fuels and toward climate solutions. For example, in 2016, the US Council on Environmental Quality released climate guidance for government agencies requiring both upstream and downstream emissions impacts to be assessed as part of their environmental analysis, in part due to public pressure to look at climate impacts in infrastructure development (Burger & Wentz, 2020).

But there is still significant scope for more research on public responses to fossil fuels and climate policy. As the editors of this volume note, fossil fuel export has been a relatively understudied dimension of research when compared to public responses to production, alternative energy, and climate change more broadly. This leaves gaps in our understanding about how shifts in public opinion or citizen mobilization may sway the policy agenda. We know little, for example, about how the general public feels about the fossil fuel production bans that have already been enacted. Has it kickstarted a necessary conversation about energy transitions or polarized the climate issue further? Does an emphasis on tangible fuels, rather than less tangible emissions, help the public better understand climate change and its causes? Can public opinion research shed light on where windows of opportunity exist to begin a conversation about phasing out fossil fuels in communities still heavily dependent on exports? And how does this policy conversation differ between low-income and high-income countries or between producers of different types of fossil fuels? Understanding these questions is critical to designing climate policies that withstand the political tailwinds facing climate action, particularly in regions reliant on fossil fuel export.

Fundamentally, climate change will not be solved by building new green infrastructure alone—to reach a true energy transition we also need to phase out the fossil fuels that are driving the climate problem (York & Bell, 2019). To achieve this, we need to better understand how the public connects the dots between fossil fuel export and climate change, and how they expect the energy transition to unfold. This will help climate policymakers design solutions that are politically palatable and effective in managing the necessary transition. Such solutions have largely evaded the climate policy world to-date but are desperately needed if we are to substantially slow the climate crisis.

References

Antón, A. (2020). Taxing crude oil: A financing alternative to mitigate climate change? *Energy Policy, 136*, 111031. https://doi.org/10.1016/j.enpol.2019.111031

Armstrong, C. (2020). Decarbonisation and world poverty: A just transition for fossil fuel exporting countries? *Political Studies, 68*(3), 671–688. https://doi.org/10.1177/0032321719868214

Bang, G., & Lahn, B. (2020). From oil as welfare to oil as risk? Norwegian petroleum resource governance and climate policy. *Climate Policy, 20*(8), 997–1009. https://doi.org/10.1080/14693062.2019.1692774

Blondeel, M., Colgan, J., & Van de Graaf, T. (2019). What drives norm success? Evidence from anti–fossil fuel campaigns. *Global Environmental Politics, 19*(4), 63–84. https://doi.org/10.1162/glep_a_00528

Bos, K., & Gupta, J. (2018). Climate change: The risks of stranded fossil fuel assets and resources to the developing world. *Third World Quarterly, 39*(3), 436–453. https://doi.org/10.1080/01436597.2017.1387477

Burger, M., & Wentz, J. (2020). Evaluating the effects of fossil fuel supply projects on greenhouse gas emissions and climate change under NEPA. *William & Mary Environmental Law and Policy Review, 44*(2), 423–530. https://scholarship.law.wm.edu/wmelpr/vol44/iss2/4.

Byrnes, R. (2020). *A global registry of fossil fuels (White paper)*. Fossil Fuel Non Proliferation Treaty. https://fossilfueltreaty.org/registry.

Carter, A. V., & McKenzie, J. (2020). Amplifying "keep it in the ground" first-movers: Toward a comparative framework. *Society & Natural Resources, 33*(11), 1339–1358. https://doi.org/10.1080/08941920.2020.1772924

Cheon, A., & Urpelainen, J. (2018). *Activism and the fossil fuel industry*. Routledge. https://doi.org/10.4324/9781351173124

Davis, S. J., Peters, G. P., & Caldeira, K. (2011). The supply chain of CO_2 emissions. *Proceedings of the National Academy of Sciences, 108*(45), 18554–18559. https://doi.org/10.1073/pnas.1107409108

Depledge, J. (2008). Striving for no: Saudi Arabia in the climate change regime. *Global Environmental Politics, 8*(4), 9–34. https://doi.org/10.1162/glep.2008.8.4.9

IPCC. (2006). 2006 IPCC guidelines for national greenhouse gas inventories. In H. Eggleston, L. Buendia, K. Miwa, T. Ngara, & K. Tanabe (Eds.), *Institute for Global Environmental Strategies (IGES) on behalf of the intergovernmental panel on climate change*. http://www.ipcc-nggip.iges.or.jp/public/2006gl/index.html.

Erickson, P., & Lazarus, M. (2013). *Accounting for greenhouse gas emissions associated with the supply of fossil fuels* (Discussion brief). Stockholm Environment Institute https://www.sei.org/publications/accounting-for-greenhouse-gas-emissions-associated-with-the-supply-of-fossil-fuels/.

Erickson, P., & Lazarus, M. (2014). Impact of the Keystone XL pipeline on global oil markets and greenhouse gas emissions. *Nature Climate Change, 4*(9), 778–781. https://doi.org/10.1038/nclimate2335

Erickson, P., Lazarus, M., & Tempest, K. (2015). *Carbon lock-in from fossil fuel supply infrastructure* (Discussion brief). Stockholm Environment Institute https://www.sei.org/publications/carbon-lock-in-from-fossil-fuel-supply-infrastructure/.

Friedman, L. (2020). *A debate pledge to 'transition' from oil puts climate at center of campaign finale.* The New York Times. https://www.nytimes.com/2020/10/23/climate/biden-debate-oil.html.

Gaulin, N., & Le Billon, P. (2020). Climate change and fossil fuel production cuts: Assessing global supply-side constraints and policy implications. *Climate Policy, 20*(8), 888–901. https://doi.org/10.1080/14693062.2020.1725409

Green, F. (2018). Anti-fossil fuel norms. *Climatic Change, 150*(1–2), 103–116. https://doi.org/10.1007/s10584-017-2134-6

Green, F., & Denniss, R. (2018). Cutting with both arms of the scissors: The economic and political case for restrictive supply-side climate policies. *Climatic Change, 150*(1–2), 73–87. https://doi.org/10.1007/s10584-018-2162-x

Green, F., & Gambhir, A. (2020). Transitional assistance policies for just, equitable and smooth low-carbon transitions: Who, what and how? *Climate Policy, 20*(8), 902–921. https://doi.org/10.1080/14693062.2019.1657379

Gupta, J. (2014). *The history of global climate governance.* Cambridge University Press. https://doi.org/10.1017/CBO9781139629072

Harrison, K. (2015). International carbon trade and domestic climate politics. *Global Environmental Politics, 15*(3), 27–48. https://doi.org/10.1162/GLEP_a_00310

Harrison, K. (2020). Political institutions and supply-side climate politics: Lessons from coal ports in Canada and the United States. *Global Environmental Politics, 20*(4), 51–72. https://doi.org/10.1162/glep_a_00579

Hazboun, S. O. (2019). A left coast 'thin green line'? Determinants of public attitudes toward fossil fuel export in the Northwestern United States. *Extractive Industries and Society, 6*(4), 1340–1349. https://doi.org/10.1016/j.exis.2019.10.009

Healy, N., Stephens, J. C., & Malin, S. A. (2019). Embodied energy injustices: Unveiling and politicizing the transboundary harms of fossil fuel extractivism and fossil fuel supply chains. *Energy Research and Social Science, 48*, 219–234. https://doi.org/10.1016/j.erss.2018.09.016

IEA. (2019). *CO_2 emissions from fuel combustion 2018.* OECD Publishing. https://doi.org/10.1787/co2_fuel-2018-en

Lazarus, M., & van Asselt, H. (2018). Fossil fuel supply and climate policy: Exploring the road less taken. *Climatic Change, 150*, 1–13. https://doi.org/10.1007/s10584-018-2266-3

Leaton, J., Ranger, N., Ward, B., Sussams, L., & Brown, M. (2013). *Unburnable carbon 2013: Wasted capital and stranded assets.* Carbon Tracker and Grantham Research Institute on Climate Change and the Environment, London School of Economics. http://www.carbontracker.org/wastedcapital.

Lee, M. (2018). Extracted carbon and Canada's international trade in fossil fuels. *Studies in Political Economy, 99*(2), 114–129. https://doi.org/10.1080/07078552.2018.1492214

Moerenhout, T., & Irschlinger, T. (2020). *Exploring the trade impacts of fossil fuel subsidies (GSI report).* International Institute for Sustainable Development. https://www.iisd.org/publications/exploring-trade-impacts-fossil-fuel-subsidies.

Muttitt, G., & Kartha, S. (2020). Equity, climate justice and fossil fuel extraction: Principles for a managed phase out. *Climate Policy, 20*(8), 1024–1042. https://doi.org/10.1080/14693062.2020.1763900

Perron, K. (2020). "Zoning out" climate change: Local land use power, fossil fuel infrastructure, and the fight against climate change. *Columbia Journal of Environmental Law, 45*(2), 573—630. https://doi.org/10.7916/cjel.v45i2.6160

Piggot, G. (2018). The influence of social movements on policies that constrain fossil fuel supply. *Climate Policy, 18*(7), 942—954. https://doi.org/10.1080/14693062.2017.1394255

Rafaty, R., Srivastav, S., & Hoops, B. (2020). Revoking coal mining permits: An economic and legal analysis. *Climate Policy, 20*(8), 980—996. https://doi.org/10.1080/14693062.2020.1719809

Ross, M. L. (2019). What do we know about export diversification in oil-producing countries? *The Extractive Industries and Society, 6*(3), 792—806. https://doi.org/10.1016/j.exis.2019.06.004

SEI, IISD, ODI, E3G, UNEP. (2020). *The production gap report: 2020 special report.* http://productiongap.org/2020report.

Seto, K. C., Davis, S. J., Mitchell, R. B., Stokes, E. C., Unruh, G., & Ürge-Vorsatz, D. (2016). Carbon lock-in: Types, causes, and policy implications. *Annual Review of Environment and Resources, 41*, 425—452. https://doi.org/10.1146/annurev-environ-110615-085934

Steininger, K. W., Lininger, C., Meyer, L. H., Munõz, P., & Schinko, T. (2016). Multiple carbon accounting to support just and effective climate policies. *Nature Climate Change, 6*(1), 35—41. https://doi.org/10.1038/nclimate2867

Temper, L., Avila, S., Bene, D. D., Gobby, J., Kosoy, N., Billon, P. L., Martinez-Alier, J., Perkins, P., Roy, B., Scheidel, A., & Walter, M. (2020). Movements shaping climate futures: A systematic mapping of protests against fossil fuel and low-carbon energy projects. *Environmental Research Letters, 15*(12), 123004. https://doi.org/10.1088/1748-9326/abc197

The White House. (2021). *President Biden takes executive actions to tackle the climate crisis at home and abroad, create jobs, and restore scientific integrity across federal government* (Fact sheet) https://www.whitehouse.gov/briefing-room/statements-releases/2021/01/27/fact-sheet-president-biden-takes-executive-actions-to-tackle-the-climate-crisis-at-home-and-abroad-create-jobs-and-restore-scientific-integrity-across-federal-government/.

Tudela, F. (2020). Obstacles and opportunities for moratoria on oil and gas exploration or extraction in Latin America and the Caribbean. *Climate Policy, 20*(8), 922—930. https://doi.org/10.1080/14693062.2020.1760772

UNEP. (2020a). *Emissions gap report 2020.* United Nations Environment Programme. https://www.unep.org/emissions-gap-report-2020.

UNEP. (2020b). *World's governments must wind down fossil fuel production by 6% per year to limit catastrophic warming* (Press release). United Nations Environment Programme https://www.unep.org/news-and-stories/press-release/worlds-governments-must-wind-down-fossil-fuel-production-6-year.

Unruh, G. C. (2000). Understanding carbon lock-in. *Energy Policy, 28*(12), 817—830. https://doi.org/10.1016/s0301-4215(00)00070-7

Widener, P. (2021). Transitioning to oil and gas exports and carbon neutrality in Aotearoa New Zealand. In S. Hazboun, & H. Schaffer Boudet (Eds.), *Public responses to fossil fuel export: Energy transition and the shifting global energy order.* Elsevier.

York, R., & Bell, S. E. (2019). Energy transitions or additions? *Energy Research & Social Science, 51*, 40—43. https://doi.org/10.1016/j.erss.2019.01.008

PART III

Public opinion on export

CHAPTER 4

The evolution of US public attitudes toward natural gas export: a pooled cross-sectional analysis of time series data (2013–2017)

Chad Zanocco[1], Shawn Hazboun[2], Greg Stelmach[3] and Hilary Boudet[3]

[1]Civil and Environmental Engineering, Stanford University, Stanford, CA, United States; [2]Graduate Program on the Environment, The Evergreen State College, Olympia, WA, United States; [3]Sociology, School of Public Policy, Oregon State University, Corvallis, OR, United States

Background

The abundance of natural gas production in North America, driven by the rapid expansion of unconventional oil and gas development, colloquially referred to as hydraulic fracturing or fracking, has transformed the US energy landscape. The United States, once a net importer of natural gas, has become a net exporter since 2019 (U.S. Energy Information Administration (EIA), 2019). Additionally, the falling cost of natural gas led to lower prices, which for consumers have translated to lower heating costs in winter, and for the electricity grid, wider usage of natural gas generation plants. While the use of natural gas for electricity production is typically considered more environmentally friendly that traditional coal-fired plants (Burnham et al., 2012; Littell, 2017), there is an ongoing debate about whether the expanded global use of natural gas presents a net positive or net negative for global carbon emissions and environmental pollution (Woollacott, 2020; Zhang et al., 2020). This debate among the environmentally minded public focuses on whether natural gas is a bridge to a lower carbon future or if it simply prolongs this transition by shifting reliance from one fossil fuel to another. And, many point out that while natural gas does not release as much carbon dioxide emissions as coal or oil, it produces methane, an even more potent greenhouse gas (Howarth, 2014).

Public Responses to Fossil Fuel Export
ISBN 978-0-12-824046-5
https://doi.org/10.1016/B978-0-12-824046-5.00007-2

The debate over natural gas has evolved over the decades (Delborne et al., 2020), and has been influenced by industry (Tabuchi, 2020), environmental groups (Kennedy, 2009; Sheppard, 2012), and policymakers, including President Obama who actively praised natural gas as an alternative to coal-fired power (Clemente, 2019). Alongside the debate over using natural gas as a source of domestic energy is the issue of whether or not the United States should export its surplus natural gas. Since the advent of unconventional natural gas extraction in the early 2000s, the United States has increasingly exported more natural gas, both via pipeline to Canada and Mexico and also on ships to other countries using liquefied natural gas (LNG) technology. The United States currently has seven LNG terminals to compress, store, and ship natural gas abroad, with four additional terminals currently under construction (Federal Energy Regulatory Commission, 2021). Yet, very little research has examined public opinion about natural gas export (Hazboun, 2019; Pierce et al., 2018) Chapter 4.

In this research, we consider how US public opinion toward the export of natural gas has evolved across time to understand changing public sentiment and develop a conceptualization of future policy support toward the export of natural gas. To do so, we focus on a time period (2013—2017) of rapid expansion of natural gas production in the United States and use multilevel regression to analyze biyearly survey data to identify temporal trends in public opinion toward natural gas export.

Natural gas: bridge fuel or a bridge too far?

Academics, policymakers, and the general public continue to debate about the long-term impacts of expanding natural gas production and consumption as a way to mitigate future climate change and its impacts (Delborne et al., 2020). Proponents of natural gas production have argued that natural gas burns much cleaner than coal or oil, and, in the context of electricity generation, is much easier to maintain in power systems where peaker power plants are needed to dynamically adjust to variability in renewable energy generation or increases in grid-level demand (Mac Kinnon et al., 2018). Additionally, the falling price of natural gas has reduced the cost of energy for many Americans, especially for those in areas that rely on gas heating during the colder seasons (Chirakijja et al., 2019). Opponents, however, argue that there are high environmental and social costs associated with using natural gas as an energy source, including environmental impacts to ecosystems during infrastructure development and operation, increased dependency of communities on extractive industries, greenhouse gas emissions, such as methane emissions related to its extraction, and CO_2 emissions related to its consumption (Alvarez et al., 2012; Howarth, 2014; Howarth et al., 2011). Yet another consideration to expanded natural gas production is its export to other international markets, which can include constructing new pipeline and facilities in or near US deepwater seaports to facilitate natural gas for export through a liquification process (LNG). Whether this infrastructure development associated with LNG presents economic opportunities or environmental benefits/costs is active area of public and academic debate (Pierce et al., 2018). Additionally, the engagement of the public in issues involving the siting of this infrastructure has been well-documented (Boudet, 2019; Boudet et al., 2018; Tran et al., 2019) for natural gas and other fossil fuel export facilities in the Pacific Northwest, with some accounts suggesting a high level of regional contention that could be viewed as a broader social movement (Cite Hazboun and Boudet Chapter 8) (Sightline Institute, 2018).

Research questions

While previous academic work has considered public opinions toward natural gas export (Hazboun, 2019; Hazboun & Boudet, 2021; Pierce et al., 2018), this work has not accounted for how public opinion around natural gas export has evolved over time. In fact, there is little knowledge about how opinion toward natural gas export has changed, especially during the 2010s, a particularly active time of LNG infrastructure proposal and development in the United States, as well as a golden decade for natural gas exploration and extraction. We therefore offer the following research question:

RQ1: How has US public opinion toward natural gas export changed across time?

Previous research has identified that supporters of natural gas export tended to be male, college educated, higher income, and politically conservative (Hazboun, 2019; Hazboun & Boudet, 2021; Pierce et al., 2018). Given that relationships between sociodemographics and opinions toward natural gas export have been previously observed in extant literature, we consider how sociodemographic characteristics either intensify, or dampen, support for natural gas export across time. Since we have no expectations about the intensity or directionality of the relationship between sociodemographics and natural gas export across time, we offer the following research question:

RQ2: What is the relationship between sociodemographic characteristics and change in opinion toward natural gas export across time?

To address these questions, we analyze a unique dataset that tracked US public opinion about natural gas export across 5 years (2013–2017).

Materials and methods

Data

In this research we apply public opinion data furnished by the University of Texas at Austin Energy Poll (UT Energy Poll), administered by Toluna, NA, a survey research firm. These public opinion data were generated via a sample drawn from a larger cohort of participants that were recruited online. Survey questions asked respondents about a variety of energy-related topics, including opinions about energy production sources and energy use behaviors, as well as sociodemographics. This poll was conducted biyearly, with survey waves typically administered in the Spring (March) and Fall (September). A question specific to natural gas export was asked on survey instruments administered from September 2013 (Wave 4) through March 2017 (Wave 12), for a total of nine survey waves. Sample size was approximately 2,000 respondents per survey wave, and when these waves were pooled, this resulted in an analytical sample of 16,805. While this survey was conducted across multiple time periods, the same respondents were not repeatedly surveyed across waves (i.e., panel design), so each wave represents a different composition of the US public (i.e., pooled cross-sectional design). Responses were weighted per wave to match population statistics from the US Census based on gender, age, education level, household income, and region. We only used this weighted data when reporting topline time trends for agreement with natural gas export; unless otherwise indicated, all data applied in this research are unweighted.

Measures

The main dependent variable applied in our analysis is agreement with natural gas export, formed from the question "To what extent do you agree or disagree with the statement below? The U.S. should permit the export of natural gas to other countries" with response categories: 1 = "Strongly disagree"; 2 = "Somewhat disagree"; 3 = "Neither"; 4 = "Somewhat agree"; 5 = "Strongly agree" (mean = 3.11; SD = 1.19). We then further dichotomized this variable as 1 = "Agree" and 0 = "Disagree or neither" (35.9% Agree) for application in regression modeling. See Fig. 4.1 for the distribution of original and recoded dependent variable formulations. We conducted analyses using both versions of these dependent variables and found that they yield similar results in analysis, so for interpretability we only applied the dichotomized version of this variable in modeling. Finally, when presenting descriptive statistics for natural gas export over time on

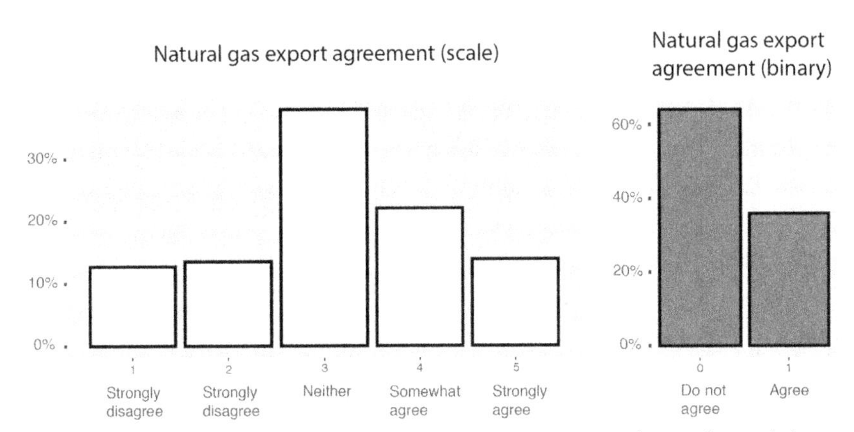

Figure 4.1 Distribution of natural gas export agreement for scale and binary measures.

weighted data, we applied a three-category version of the dependent variable ("Disagree"; "Neither"; "Agree") so that changes in the "Neither" response category can be easily tracked across time.

For sociodemographic variables we included measures of age, gender, race, education, income, and political party affiliation (see Table 4.1). Age is reported on an 11-point scale from 1 = "18–24 years" to 11 = "75 years or older" (median = 6 or "45–49 years"), gender as male versus female (47% male), race/ethnicity as white versus nonwhite (81% white), education as less than bachelor's degree versus bachelor's degree or higher (41% bachelor's or higher), and household income is represented on an 8-point scale from 1 = "Less than $20,000" to 8 = "$200,000 or more" (median = 4 or "$40,000 to less than $50,000"). Political party affiliation responses resulted in the following categories: Democrat (42%), Republican (31%), Libertarian (4%), and Independent/other (22.1%). Since Democratic affiliation comprises the largest percentage of our respondents, we use this as the reference category in analysis.

Finally, we consider multiple measures of time related to when the survey was administered. Survey waves were conducted biannually, spaced approximately six months apart (March, September) with one exception in 2016 when the survey was administered in January. Due to these regular intervals, we use survey wave administration periods as a representation of nine sequential sub-yearly time periods. We also combine these wave time periods into yearly measures representing five years from 2013 to 2017. Both of these time measures are included throughout our analysis.

Table 4.1 Description of measures.

Measure	Description/question wording	Summary
Agreement with natural gas export	Question: The US should permit the export of natural gas to other countries	Range: 1 = "Strongly disagree" to 5 = "Strongly agree"; Mean = 3.109; Median = 3 or "Neither"; Std. dev. = 1.186
Age	Question: In which of the following age groups do you fall?	Range: 1 = "18–24 years" to 11 = "75 or older"; Mean = 5.797; Median = 6 or "45–49 years"; Std. dev. = 3.114
Gender	Question: Are you ... ?	47.1% male
Race/ ethnicity	Question: What is your racial or ethnic heritage?	81.1% white
Education	Question: What is the last grade of school you completed?	41.0% bachelor's degree or higher
Income	Question: Which of the following income groups includes your total family income before taxes?	Range: 1 = "Less than $20,000" to 8 = "$200,000 or more"; Mean = 3.966; Median = 4 or $40,000 to less than $50,000; Std. dev. = 2.034
Political affiliation	Question: Generally speaking, which of the following best describes your political affiliation?	42.9% Democrat; 31.1% Republican 22.1%; Independent/other; 3.9% Libertarian
Year	Year survey was administered	Range: 2013–2017; n = 5
Waves	Survey waves	n = 9
ZIP codes	Five-digit ZIP code	n = 8124
States	US states and Washington, D.C.	n = 51

Analysis

The primary analytical approach that we apply to our data is multilevel regression modeling. Multilevel regression analysis has been leveraged in a variety of research settings to improve the efficiency of estimation when contextual or place-based characteristics influence phenomena of interest (Gelman, 2006). Multilevel modeling is being increasingly applied in research where space and geographic settings are expected to impact public opinion (Zanocco et al., 2018; Zanocco, Boudet, Clarke, & Howe, 2019;

Zanocco et al., 2019), and, in a particular, has been used to model how physical proximity toward energy development shapes attitudes toward said development (Zanocco et al., 2018; Zanocco et al., 2019; Zanocco et al., 2020).

All multilevel models were estimated using the R package lme4 (Bates, 2014). We apply binary logistic multilevel regression using binary natural gas export agreement as the dependent variable (0 = Disagree or neither; 1 = Agree), sociodemographic characteristics of the survey respondent and associated time measures modeled as lower-level variables (i.e., fixed effects), and contextual characteristics including ZIP code and state modeled as higher-level variables (i.e., random effects). As the data applied in this analysis are a collection of cross sections of the US public, rather than repeated observations of the same individual across time (i.e., longitudinal or panel analysis), we utilize a pooled cross-sectional time series approach where we pool all respondents in a single analysis and include a measure of time as a fixed effect in our model specifications. These time effects are applied in modeling using two analytical approaches, first by including year or wave as a continuous time measure so we can test for linear time trends, then by including a series of binary time variables in order to control for the effects associated with individual years/waves, using the first year/wave as the reference category. Such a modeling approach also allows us to model moderation effects by including interaction terms between our time measures and sociodemographic characteristics.

Findings

Patterns in agreement across time

We first consider how respondent agreement toward natural gas export has changed across time. For this analysis, we utilized a three-category version of the dependent variable ("Disagree," "Neither," "Agree") for yearly and subyearly time measures, displaying the weighted average of respondents per timestep (Fig. 4.2). In Fig. 4.2, we see a pattern that descriptively suggests that agreement for US natural gas export is increasing across time, disagreement with US natural gas export is decreasing across time, and a neutral opinion toward natural gas export remains stable. For example, in the year 2013 the sampled respondents report that Disagree is 32.7%, Neither is 37.1%, and Agree is 30.1%, compared to 2017 where respondents report that Disagree is 20.1%, Neither is 38.0%, and Agree is 41.9%. This same pattern is observed for the biannual survey wave measure, where the mean for each wave in temporal sequence is higher than the previous survey wave.

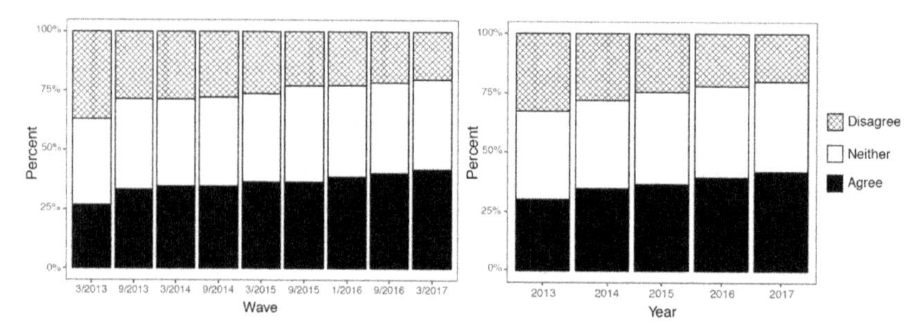

Figure 4.2 Agreement with US natural gas export across time by survey wave (left) and by year (right). Categories are weighted to match the US Census.

Modeling agreement with natural gas export

We next consider binary logistic multilevel regression models predicting agreement with natural gas export using sociodemographics and time measures as fixed effects and participant ZIP code and state as random effects. All model results are reported as an odds ratio (OR) (as a reminder, an odds ratio of 1 indicates no association, larger than 1 a positive association and less than 1 a negative association).

The relationship between individual-level characteristics on agreement with natural gas export is stable across our models using alternative formulations of time (Table 4.2: Models 1–3). Those who are older (OR = 0.944; $P < .001$) and those who self-identified as being Independent/other (vs. Democrats; OR = 0.808; $P < .001$) are less likely to agree with natural gas export, while males (vs. other; OR = 2.224; $P < .001$), those with bachelor's degrees (OR = 1.561; $P < .001$) and higher incomes (OR = 1.095; $P < .001$) have higher levels of agreement (Table 4.2: Model 1). Republican (OR = 1.117; $P < .05$) and Libertarian (OR = 1.294; $P < .01$) party affiliations are also associated with more agreement with natural gas export compared to the reference category of Democrat (Table 4.2: Model 1). We do not find statistically significant relationships ($P < .05$) related to race/ethnicity (white vs. nonwhite) across any of the models.

Consistent with the temporally varying pattern that we observe in the previous figures displaying change in natural gas agreement over time (Fig. 4.2), we find that all included time effects are statistically significant in our analytical modeling. First, we test years modeled as a linear trend (continuous measure) and find that is significant and positive, suggesting an increase in agreement across time (Model 1; OR = 1.101; $P < .001$).

Table 4.2 Binary logistic multilevel regression models predicting agreement with natural gas export.

	Agree with natural gas export					
	Model 1		**Model 2**		**Model 3**	
	Odds Ratios	*P-value*	*Odds Ratios*	*P-value*	*Odds Ratios*	*P-value*
Age	0.944***	<.001	0.944***	<.001	0.944***	<.001
Male (vs. female)	2.224***	<.001	2.220***	<.001	2.216***	<.001
White (vs. nonwhite)	0.96	.406	0.96	.395	0.96	.429
Bachelor's or higher	1.561***	<.001	1.558***	<.001	1.562***	<.001
Income	1.095***	<.001	1.095***	<.001	1.097***	<.001
Republican (vs. democrat)	1.117*	.011	1.120**	.009	1.113*	.014
Libertarian (vs. democrat)	1.294**	.002	1.292**	.003	1.279**	.004
Independent/other (vs. democrat)	0.808***	<.001	0.809***	<.0001	0.803***	<.001
Years (2013–2107)	1.101***	<.001				
Waves (1–9)			1.052***	<.001		
Year: 2014 (vs. 2013)					1.232***	<.001
Year: 2015 (vs. 2013)					1.316***	<.001
Year: 2016 (vs. 2013)					1.424***	<.001
Year: 2017 (vs. 2013)					1.438***	<.001
(Intercept)	0.205***	<.001	0.198***	<.001	0.234***	<.001
Random effects	*variance*		*variance*		*variance*	
ZIP codes (intercept)	0.34		0.34		0.34	
State (intercept)	0.01		0.01		0.01	
Akaike Information Criterion	20760.08		20756.42		20758.4	
N	16805		16805		16805	

Significance levels: * $P < .05$; ** $P < .01$; *** $P < .001$.

This same increasing trend is observed for waves in Model 2 (OR = 1.052; $P < .001$). Additionally, in Model 3, instead of including a linear time trend we include yearly time dummies for 2014–2017, with 2013 as the reference category. Each of these time dummies is statistically significantly, have higher agreement than the baseline year of 2013, and are increasing in magnitude each year.

When predicted probabilities for years (2013–2017) in Model 1 are plotted, the presence of this increasing time trend is further reinforced, demonstrated by the positive relationship between agreement with natural gas export and years (Fig. 4.3). Holding all other model covariates constant, public agreement increases on average by approximately 2.2% each year from 2013 to 2017.

We next consider how our modeled sociodemographics may be related to this change in agreement with natural gas export by interacting each statistically significant sociodemographic characteristic from Table 4.2 with the yearly time trends, with different interaction terms applied in separate models (Table 4.3). The following sociodemographic characteristics interacted with time are statistically significant in these time interaction models: age (OR = 0.985; $P < .01$); income (OR = 1.016; $P < .05$); and bachelor's degree or (OR = 1.073; $P < .05$). Higher income, younger age, and college education are all associated increased agreement in natural gas export across time. See Fig. 4.4 for interaction plots displaying these patterns across time. We do not, however, find a moderation effect associated with male (vs. female) or political affiliation measures.

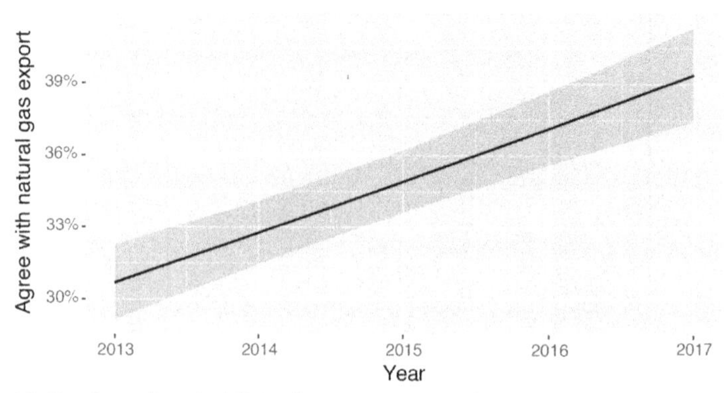

Figure 4.3 Predicated probabilities for agreement with natural gas export in response to time (years 2013–2017). Estimates are derived from Model 1.

Table 4.3 Binary logistic multilevel regression models predicting agreement with natural gas export with interaction effects for statistically significant ($P < .05$) sociodemographic factors.

	Agree with natural gas export									
	Model 4		**Model 5**		**Model 6**		**Model 7**		**Model 8**	
	Odds Ratios	*P-value*	*Odds Ratios*	*P-value*	*Odds Ratios*	*P-value*	*Odds Ratios*	*P-value*	*Odds Ratios*	*P-value*
Age	0.999	.962	0.945***	<.001	0.945***	<.001	0.945***	<.001	0.945***	<.001
Male (vs. female)	2.224***	<.001	1.934***	<.001	2.220***	<.001	2.220***	<.001	2.224***	<.001
White (vs. nonwhite)	0.953	.321	0.959	.406	0.955	.429	0.95	.429	0.958	.378
Bachelor's or higher	1.556***	<.001	1.560***	<.001	1.199***	<.001	1.555***	<.001	1.560***	<.001
Income	1.093***	<.001	1.095***	<.001	1.094***	<.001	1.032***	<.001	1.095***	<.001
Republican (vs. democrat)	1.116*	.012	1.118*	.011	1.119*	.014	1.122*	.014	1.135	.318
Libertarian (vs. democrat)	1.292**	.003	1.296**	.002	1.295**	.004	1.295**	.004	1.115	.68
Independent/other (vs. democrat)	0.810***	<.001	0.808***	<.001	0.808***	<.001	0.810***	<.001	0.917	.547
Years (2013–2107)	1.201***	<.001	1.08***	<.001	1.064**	.001	1.03	.355	1.108***	<.001
Age × Years	0.985**	.001								
Male (vs. female) × Years			1.037	.174						
Bachelor's or higher × Years					1.073*	.010				

Continued

Table 4.3 Binary logistic multilevel regression models predicting agreement with natural gas export with interaction effects for statistically significant ($P < .05$) sociodemographic factors.—cont'd

	Agree with natural gas export									
	Model 4		**Model 5**		**Model 6**		**Model 7**		**Model 8**	
	Odds Ratios	*P-value*	*Odds Ratios*	*P-value*	*Odds Ratios*	*P-value*	*Odds Ratios*	*P-value*	*Odds Ratios*	*P-value*
Income × Years							1.016*	.019		
Republican (vs. democrat) × Years									0.996	.904
Libertarian (vs. democrat) × Years									1.039	.553
Independent/other (vs. democrat) × Years									0.966	.35
(Intercept)	0.149***	<.001	0.221***	<.001	0.234***	<.001	0.265***	<.001	0.201***	<.001
Random effects	*variance*		*variance*		*variance*		*variance*		*variance*	
ZIP codes (intercept)	0.34		0.34		0.34		0.34		0.34	
State (intercept)	0.01		0.01		0.01		0.01		0.01	
Akaike Information Criterion	20750.28		20760.05		20755.49		20756.65		20764.6	
N	16805		16805		16805		16805		16805	

Significance levels: * $P < .05$; ** $P < .01$; *** $P < .001$.

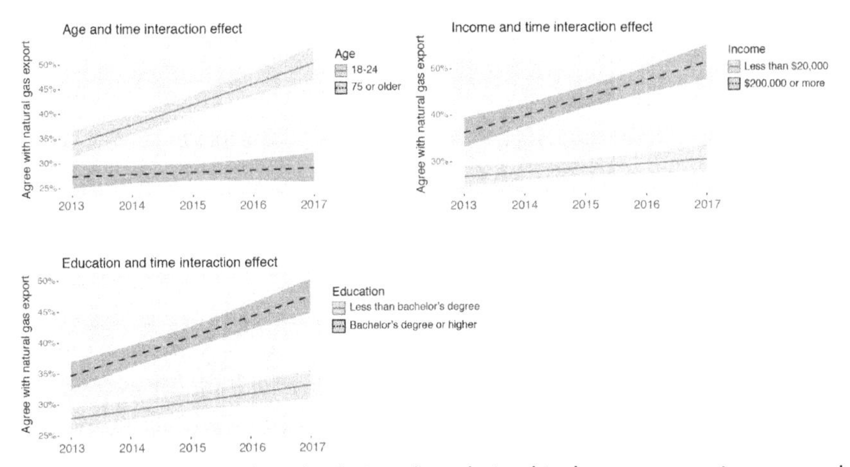

Figure 4.4 Interaction plots displaying the relationship between age, income, and education in response to time (years 2013–2017). For age and income, which are measured on a scale, the interaction effect for lowest and highest end of the scale are displayed for visualization purposes.

Discussion

We find strong evidence that sociodemographic characteristics are related to attitudes toward US natural export, that natural gas export opinion changed across the 2013–2017 time period, and that some sociodemographic characteristics are related to these to these temporal trends. Consistent with existing research on opinions toward natural gas export (Pierce et al., 2018), we find that gender, income, education, age, and political affiliation are related to agreement with natural gas export. However, our unique dataset also allows us to explore how public opinion toward natural gas export is changing across time. Surprisingly, we observe a robust and consistent time trend where, in both yearly and subyearly analysis, average overall agreement increases for natural gas export. We further explore this time trend and find that sociodemographic characteristics including age, income, and education are associated with intensified acceptance of natural gas export over time. However, we do not observe that political affiliation is related to this time trend, suggesting that political affiliation did not play a strong a role in changing opinion across time. Rather, political affiliation appears to either have a more static association with agreement toward natural gas export or the composition of people within these political affiliation groups could be changing across time.

As such, demographic characteristics such as age, education, and income may provide a more stable temporal measurement that is better suited for pooled cross-sectional time series analysis.

While it is challenging to project these findings into the future, given the rapidly evolving energy landscape, as well as changing US perspectives toward energy security, the environment, and the economy, our findings identify an upward trend in acceptance of natural gas export that appears quite durable. Yet, specific cases of opposition to the siting of LNG export facilities and associated pipeline infrastructure since this survey was conducted may have already begun to move the needle against this public sentiment. With these substantial caveats noted, if we take these findings and project them to 2021, for example, we would predict that more than half (50.7%) of the US public would be in agreement with natural gas export. Though the trend we have identified likely does not have such a linear relationship that could be projected into the future in this way, more recent public opinion work suggests the public is increasingly supportive of exporting natural gas (Hazboun, 2019; Pierce et al., 2018).

Some limitations to our study are worth noting. First, each time period included in the survey is a different cross section of the US population, so at each timestep we are comparing a different sample of respondents. In this respect, we are only able to make comparisons across individuals at different timesteps, and thus have no insights into how the trend we have observed relates to a change in opinion at the individual level. A panel design, where the same respondent is surveyed at different time periods, would allow for this type of insight. While a panel design of the length of time we present here (five years) would likely be infeasible due to prohibitive costs and participant attrition, a shorter panel, covering two years, could be feasible. Second, it's worth noting that our sample was not probability based—instead, it constitutes an online convenience sample, and as such our ability to make inferences to the general public are limited. Future research should keep these considerations in mind when seeking to understand how perspectives toward natural gas export may be changing, or remaining the same, across new and expanded time horizons.

References

Alvarez, R. A., Pacala, S. W., Winebrake, J. J., Chameides, W. L., & Hamburg, S. P. (2012). Greater focus needed on methane leakage from natural gas infrastructure. *Proceedings of the National Academy of Sciences, 109*(17), 6435–6440. https://doi.org/10.1073/pnas.1202407109

Bates, et al. (2014). https://arxiv.org/abs/1406.5823.
Bates, D., Machler, M., Bolker, B., Walker, S., et al. (2015). Fitting linear mixed-effects models using lme4. *Journal of Statistical Software, 67*. https://doi.org/10.18637/jss.v067.i01
Boudet, H. S. (2019). Public perceptions of and responses to new energy technologies. *Nature Energy, 4*(6), 446–455. https://doi.org/10.1038/s41560-019-0399-x
Boudet, H. S., Zanocco, C. M., Howe, P. D., & Clarke, C. E. (2018). The effect of geographic proximity to unconventional oil and gas development on public support for hydraulic fracturing. *Risk Analysis, 38*(9), 1871–1890. https://doi.org/10.1111/risa.12989
Burnham, A., Han, J., Clark, C. E., Wang, M., Dunn, J. B., & Palou-Rivera, I. (2012). Life-cycle greenhouse gas emissions of shale gas, natural gas, coal, and petroleum. *Environmental Science and Technology, 46*(2), 619–627. https://doi.org/10.1021/es201942m
Chirakijja, J., Jayachandran, S., & Ong, P. (2019). *Inexpensive heating reduces winter mortality*. Retrieved from https://www.nber.org/papers/w25681.
Clemente, J. (2019). *President Obama's support for America's shale oil and natural gas*. Forbes. https://www.forbes.com/sites/judeclemente/2020/12/31/president-obamas-support-for-americas-shale-oil-and-natural-gas/.
Delborne, J. A., Hasala, D., Wigner, A., & Kinchy, A. (2020). Dueling metaphors, fueling futures: "Bridge fuel" visions of coal and natural gas in the United States. *Energy Research and Social Science, 61*, 101350. https://doi.org/10.1016/j.erss.2019.101350
Federal Energy Regulatory Commission. (2021). *North American LNG export terminals – existing, approved not yet built, and proposed*. Federal Energy Regulatory Commission. Retrieved from https://cms.ferc.gov/media/north-american-lng-export-terminals-existing-approved-not-yet-built-and-proposed-1.
Gelman, A. (2006). Multilevel (hierarchical) modeling: What it can and cannot do. *Technometrics, 48*(3), 432–435. https://doi.org/10.1198/004017005000000661
Hazboun, S. O. (2019). A left coast 'thin green line'? Determinants of public attitudes toward fossil fuel export in the northwestern United States. *The Extractive Industries and Society, 6*(4), 1340–1349. https://doi.org/10.1016/j.exis.2019.10.009
Hazboun, S. O., & Boudet, H. S. (2021). Natural gas – friend or foe of the environment? Evaluating the framing contest over natural gas through a public opinion survey in the Pacific Northwest. *Environmental Sociology, 0*(0), 1–14. https://doi.org/10.1080/23251042.2021.1904535
Howarth, R. W. (2014). A bridge to nowhere: Methane emissions and the greenhouse gas footprint of natural gas. *Energy Science and Engineering, 2*(2), 47–60. https://doi.org/10.1002/ese3.35
Howarth, R. W., Santoro, R., & Ingraffea, A. (2011). Methane and the greenhouse-gas footprint of natural gas from shale formations. *Climatic Change, 106*(4), 679. https://doi.org/10.1007/s10584-011-0061-5
Kennedy, R., Jr. (2009). *How to end America's deadly coal addition*. Financial times. Retrieved from https://www.ft.com/content/58ec3258-748b-11de-8ad5-00144feabdc0#axzz1qaQSM4kw.
Littell, D. (2017). Natural gas: Bridge or wall in transition to low-carbon economy? *Natural Gas & Electricity, 33*(6), 1–8. https://doi.org/10.1002/gas.21953
Mac Kinnon, M. A., Brouwer, J., & Samuelsen, S. (2018). The role of natural gas and its infrastructure in mitigating greenhouse gas emissions, improving regional air quality, and renewable resource integration. *Progress in Energy and Combustion Science, 64*, 62–92. https://doi.org/10.1016/j.pecs.2017.10.002
Pierce, J. J., Boudet, H., Zanocco, C., & Hillyard, M. (2018). Analyzing the factors that influence U.S. public support for exporting natural gas. *Energy Policy, 120*, 666–674. https://doi.org/10.1016/j.enpol.2018.05.066

Sheppard, K. (2012). *Natural gas puts greens in tough spot.* Mother Jones. Retrieved from https://www.motherjones.com/environment/2012/09/natural-gas-fracking-sierra-nrdc/.

Sightline Institute. (2018). *The thin green line.* Sightline Institute. Retrieved from https://www.sightline.org/research/thin-green-line/.

Tabuchi (2020). https://www.nytimes.com/2020/11/11/climate/fti-consulting.html.

Tran, T., Taylor, C. L., Boudet, H. S., Baker, K., & Peterson, H. L. (2019). Using concepts from the study of social movements to understand community response to liquefied natural gas development in Clatsop county, Oregon. *Case Studies in the Environment.* https://doi.org/10.1525/cse.2018.001800

U.S. Energy Information Administration (EIA). (2019). *United States has been a net exporter of natural gas for more than 12 consecutive months.* Today in Energy. Retrieved from https://www.eia.gov/todayinenergy/detail.php?id=39312#.

Woollacott, J. (2020). A bridge too far? The role of natural gas electricity generation in US climate policy. *Energy Policy, 147*, 111867. https://doi.org/10.1016/j.enpol.2020.111867

Zanocco, C., Boudet, H., Clarke, C. E., & Howe, P. D. (2019). Spatial discontinuities in support for hydraulic fracturing: Searching for a 'Goldilocks Zone'. *Society & Natural Resources, 32*(9), 1065–1072. https://doi.org/10.1080/08941920.2019.1616864

Zanocco, C., Boudet, H., Nilson, R., & Flora, J. (2019). Personal harm and support for climate change mitigation policies: Evidence from 10 U.S. communities impacted by extreme weather. *Global Environmental Change, 59*, 101984. https://doi.org/10.1016/j.gloenvcha.2019.101984

Zanocco, C., Boudet, H., Nilson, R., Satein, H., Whitley, H., & Flora, J. (2018). Place, proximity, and perceived harm: Extreme weather events and views about climate change. *Climatic Change, 149*(3), 349–365. https://doi.org/10.1007/s10584-018-2251-x

Zanocco, C., Boudet, H., Clarke, C. E., Stedman, R., & Evensen, D. (2020). NIMBY, YIMBY, or something else? Geographies of public perceptions of shale gas development in the Marcellus shale. *Environmental Research Letters.* https://doi.org/10.1088/1748-9326/ab7d01

Zhang, B., Mildenberger, M., Howe, P. D., Marlon, J., Rosenthal, S. A., & Leiserowitz, A. (2020). Quota sampling using Facebook advertisements. *Political Science Research and Methods*, 1–7. https://doi.org/10.1017/psrm.2018.49

CHAPTER 5

Drivers of US regulatory preferences for natural gas export

Greg Stelmach[1], Jonathan Pierce[2], Chad Zanocco[3] and Hilary Boudet[1]
[1]Sociology, School of Public Policy, Oregon State University, Corvallis, OR, United States; [2]University of Colorado Denver, School of Public Affairs, Denver, CO, United States; [3]Civil and Environmental Engineering, Stanford University, Stanford, CA, United States

Introduction

Natural gas extraction, particularly through fracking, has been the subject of extensive research in the past few years, but far less attention has been paid to what happens to natural gas once it is out of the ground. The dramatic rise of US natural gas production has prompted significant investment in exporting infrastructure, particularly in the form of liquefied natural gas (LNG). Export of shale gas in the form of LNG from the continental United States did not begin until 2016 (Gronholt-Pedersen, 2016) but has since become a major industry. Within the United States, natural gas is usually transported via pipelines. However, to get natural gas from the United States to markets in other countries in Asia or Europe, it is liquefied. LNG is natural gas that has been cooled into a liquid state at about −260°F, often at liquefication units located near export terminals. From the export terminals, LNG is shipped using LNG carriers to import terminals around the world (U.S. EIA, 2019a,b,c).

In 2019, the United States was the third largest exporter of LNG in the world, behind only Qatar and Australia. From 2018 to 2019 alone, the amount of LNG export increased by 61% (U.S. EIA, 2019a). Industry has been proposing and building new liquefication units and export terminals, with two new liquefication units and export terminals coming into service in Texas and Louisiana in 2019. As of September 2020, there were six liquefication terminals for LNG export in the United States. Four of them are in the Gulf of Mexico (Sabine Pass, LA; Corpus Christi, TX; Cameron, LA; and Freeport, TX); one terminal in Cove Point, MD; and a sixth liquefication plant and export terminal at Elba Island, GA (FERC, 2020).

Public Responses to Fossil Fuel Export
ISBN 978-0-12-824046-5
https://doi.org/10.1016/B978-0-12-824046-5.00002-3

An additional five terminals are approved and under construction, all in Louisiana and Texas. Based on this increased capacity, the US EIA projects that the United States will be the largest exporter of natural gas in the world by 2025 (U.S. EIA, 2019b).

The United States exports LNG primarily to Asian and European markets, with Asian exports surpassing those to Europe in 2020 (U.S. EIA, 2021). The primary countries in Asia the United States exports to are South Korea, Japan, and China; in Europe, exports primarily go to Spain, the United Kingdom, and Turkey. The United States also exports LNG to Mexico, Chile, and Brazil as well (U.S. EIA, 2021). The largest importer of natural gas in the world is China, recently surpassing Japan and Germany. China is seeking to shift its primary source of energy from coal to natural gas thus increasing demand. However, LNG exports from the United States only accounted for 7% of China's total imports of LNG during the first 6 months of 2018 prior to the placement of tariffs upon LNG imports from the United States (U.S. EIA, 2019a). Given the significant trade implications, LNG export intersects a diverse range of policy spheres, such as geopolitics, foreign policy, and macroeconomic trade policy.

The export as well as import of natural gas in the United States is regulated by the Department of Energy (DOE) as per the Natural Gas Act of 1938. The DOE, among other responsibilities, must approve any proposal to export LNG to a country without a free trade agreement with the United States, and such a determination must find that the sale is in the public interest (Chanis, 2012). The Federal Energy Regulatory Commission (FERC) is responsible for authorizing the siting and construction of onshore and nearshore LNG import and export facilities in the United States. As part of this process, FERC prepares environmental assessments or impact statements for proposed LNG facilities, in the process addressing potential environmental and socioeconomic impacts. As export terminals are frequently sited at deep-water ports, the FERC process considers the effects of any dredging or other construction impacts on the local ecology or fisheries. Additionally, new terminals may require the construction of a pipeline to connect to existing infrastructure, a process that can ultimately involve the exercise of eminent domain to route the pipeline through privately held land. This process is also controversial with respect to global environmental impacts, as recently the DOE found that US LNG exports for power production in Europe and Asia will not increase greenhouse gas emissions from a lifecycle perspective in comparison to regional coal extraction. This finding by the Trump administration is in agreement with the previous analysis done by the Obama administration (Klump, 2019a,b). Industry leaders have praised this finding and agree that LNG exports address climate change while also helping national security and local economics (Klump, 2019b). In contrast, environmental advocacy groups such as the Sierra Club argue that the DOE underestimated emissions from natural gas and that while emissions may be better than coal, the DOE did not consider how the export of LNG will displace renewable sources of energy (Klump, 2019a).

Given the many policy spheres that overlap with LNG export, regulation is a multipronged process that must consider a variety of dimensions. However, little is known about how the US public views LNG export regulation. This chapter aims to demonstrate the current knowledge gaps about public attitudes toward LNG exporting, formulate hypothesized relationships based on comparative industries, test these associations using a national dataset, and lastly discuss the implications for policymakers.

Literature review

While research into natural gas in the United States focuses much attention on fracking, limited scholarship exists on LNG export, particularly with regard to public attitudes toward its regulation. Where it has been studied, most LNG export research has focused on macroeconomic considerations (Bernstein et al., 2016), climate impacts (Gilbert & Sovacool, 2017), and geopolitical implications (Medlock et al., 2014). Though these works do not directly address public attitudes of LNG export, they do support the notion that regulation covers a wide array of interests. Pierce, Boudet, Zanocco, and Hillyard (2018) previously used the same survey data we explore here to examine support/opposition toward natural gas export but not attitudes toward regulation. Chanis (2012) noted that the issue of LNG exports is problematic in the United States because of the many competing and contradictory constituencies, such as industry, environmental groups, labor unions, and national security groups. As such, our analysis of public policy preferences for regulation incorporates a diverse range of factors.

Recent research examining the framing of natural gas in the public sphere in the United States found that, while natural gas has been framed as a bridge fuel to a renewable energy future, increasingly this frame has been contested by actors who see reliance on natural gas as hindering a transition to renewables (Delborne et al., 2020). At the same time, surveys on public perceptions of natural gas compared to other fossil fuels show that the public tends to view natural gas more favorably than other fossil fuels but less so than renewables (Ansolabehere & Konisky, 2014; Hazboun & Boudet, 2020). In particular, men, conservatives, and those who did not believe global warming is human caused were more likely to view natural gas as environmentally friendly. Such views of natural gas also translated into more support for its use in electricity generation and for export (Ansolabehere & Konisky, 2014; Hazboun & Boudet, 2020).

Existing scholarship provides an important foundation for understanding the policy and regulatory dimensions of LNG export, yet there is a significant knowledge gap in understanding how the US public views export and its regulation. We draw on the more robust literature concerning attitudes about related energy issues, such as fracking, power plant siting, natural gas, and emerging technologies (e.g., nanotechnology), to provide a theoretical framework of factors that would be reasonably expected to influence attitudes about LNG export regulation. Additionally, we consider literature more broadly focused on climate change that has specific consideration of regulatory policy preferences, to consider a perspective more centered on the notion of regulation instead of general attitude. Overall, we highlight five factors of primary concern for our exploratory analysis.

Age

Literature on age and attitudes to new energy policy has been somewhat mixed. A study of public attitudes toward the construction of new power plants by Ansolabehere and Konisky (2009) found that older respondents were more supportive of new natural gas power plants. Gravelle and Lachapelle (2015) found a similar association with support for building the Keystone XL pipeline, as did Clarke et al. (2016) when studying support for unconventional oil and natural gas development. Davis and Fisk (2014), on the other hand, found no significant relationship between respondent age and support or opposition to fracking regulation. Dietz et al. (2007) found older age not only correlated with more support for climate change reduction policies but also correlated with higher trust in industry actors, which might indicate support for less regulation. The only significant polling conducted related to LNG export has found that older Americans are generally less supportive of LNG export (University of Texas Energy Poll, 2013). Given the lack of consensus, we pose the following research question:

RQ_1: Does age have a significant association with support for LNG export regulatory policy?

Gender

Existing research on the gender divide on energy regulation overwhelmingly finds that women are opposed to energy policy that is harmful to the environment. Pierce et al. (2018) and Hazboun and Boudet (2021) found that supporters of LNG export tended to be more male than female. Other literature shows that women have higher levels of opposition than men to new natural gas power plants (Ansolabehere & Konisky, 2014; Hazboun & Boudet, 2020), fracking (Boudet et al., 2014; Clarke et al., 2016), and pipelines (Gravelle & Lachapelle, 2015). Women also have been shown to support higher levels of regulation of fracking (Davis & Fisk, 2014) and be more supportive of climate change policy (Dietz et al., 2007). A potential explanation for the difference is in risk perception, as Flynn et al. (1994) found that across 25 hypothetical hazards, women perceived significantly higher risk than men in each case. As such, we ask the following:

RQ_2: Do female respondents support higher levels of LNG export regulation than male respondents?

Political ideology

Similar to the literature on gender, research examining the relationship of political ideology has consistently found a partisan divide when it comes to energy issues. Individuals that identify as Republicans have been found to be more supportive of fracking (Clarke et al., 2016) and building new natural gas power plants (Ansolabehere & Konisky, 2009). Those identifying as Democrats have also been found to be more likely to support drilling regulations (Davis & Fisk, 2014). The University of Texas Energy Poll (2013), however, found that both Democrats and Republicans had similar levels of support for LNG export, with both being more favorable than independents. Hazboun (Hazboun & Boudet, 2021) found in a survey of Washington residents that liberal respondents were less likely to support exporting gas, coal, and oil. Individuals that identify as politically conservative have likewise been found to be more supportive of fracking (Boudet et al., 2014) and pipelines (Gravelle & Lachapelle, 2015). Evensen and Stedman (2017) conducted both national and regional (Marcellus Shale) surveys, finding that conservatives were expected to be more supportive of general shale development at both scales, though there was a stronger association in the Marcellus Shale survey. Finally, Malin et al. (2017) found that free market ideology—a belief commonly associated with political conservatism—correlated with desire for less federal regulation of unconventional oil and gas extraction. As such, we pose our next research question:

RQ3: Are politically conservative respondents less supportive of LNG export regulation?

Awareness

There is little consensus on whether or not an individual's self-reported awareness or familiarity with a sector of the energy industry has a significant association with support or opposition. Boudet et al. (2014) found a positive association between familiarity and opposition to fracking, but Stedman et al. (2016) found, in a comparison of US and UK public perceptions of fracking, that American respondents' reported knowledge was unrelated to their support or opposition. And Hazboun (2019) found that increased familiarity correlated with support for exporting natural gas, coal, and oil. Kahan et al. (2009) examined the role of familiarity in a study of public support for nanotechnology (another emerging technology) and found that the "familiarity hypothesis," the idea that support will grow as awareness expands, did not hold on its own. Instead, cultural cognition, which relate to an individual's worldviews (hierarchical, egalitarian, individualistic, communitarian), was the determining factor in predicting respondents' perceptions of the net benefits and risks. Because there is no consistent overall trend in the literature from which to predict an association, we ask:

RQ4: Is there a significant association between awareness of LNG export and support for increased regulation?

Benefit and risk perception

As discussed in the preceding section, the export of natural gas involves a number of benefits and risks. Davis (2012) highlighted the various energy security, economic, and political benefits of increased fracking, as well as the potentially harmful environmental and public health consequences. However, the connection between impacts and desire for regulation is not always made. Malin et al. (2017) found that Colorado respondents who perceived more negative impacts from unconventional oil and gas extraction also supported federal deregulation for the industry. In other similar energy policy areas, perceived benefits and risks of an activity have been found to correlate with attitude toward that activity. For example, Clarke et al. (2016) found a significant relationship between a respondent's perceived net risk/benefit and support or opposition for unconventional oil and natural gas development. Similarly, perceived environmental harm and adverse impacts on energy costs have had significant association with respondents opposing building new natural gas power plants (Ansolabehere & Konisky, 2009). Further, it is important to consider benefit and risk perceptions as an intermediary in regulation preferences. Multiple studies referenced in preceding sections (Flynn et al., 1994; Kahan et al., 2009) indicate that factors such as gender and cultural cognition correlate to risk and benefit perceptions, which in turn correlate with overall preferences. Therefore, we considered benefit and risk perceptions as both independent and dependent variables to better understand their roles in respondent views about LNG regulation. Our final research question poses the following:

RQ_5: Are there significant associations between respondent perceptions of LNG export benefits and risks and support for LNG export regulation?

Data and methods

The data used for this analysis come from an online survey administered through Amazon Mechanical Turks (MTurk), an online crowdsourcing platform where individuals complete specified tasks for monetary compensation. MTurk is an increasingly common recruitment method for public opinion research (McCright et al., 2013). Research has found that samples drawn from MTurk to be more representative of US populations than in-person convenience samples, but less representative than probability-based national samples (Berinsky et al., 2012). Therefore, our sample—while limited in terms of external validity—is sufficient to explore how the factors listed above track with support for the regulation of natural gas export. In total, we received 1042 responses from participants all over the United States across a 24-hour period on June 28, 2017. Respondents were compensated $0.50 for completing the survey, which took approximately 5 min.[1] We did not exclude any of our 1042 responses from the data set. Excluding subjects poses a tradeoff between improved data quality and reduced sample representativeness. We erred on the side of sample representativeness as this was exploratory and we wanted to maximize national representativeness. The relevant variables used in this analysis are explained below, along with tables of summary statistics.

Regulatory Policy Preference. The measure of regulatory policy preferences was constructed from a total of nine survey items that asked respondents to indicate the extent to which the DOE should consider a variety of factors when it regulates LNG export, on a scale of "Not at all" (1) to "A great deal" (5). As already discussed, LNG export has many dimensions that could be regulated, including economic, environmental, geopolitical, and public health implications (Adams, 2014; Bousky & Harrison, 2016). This led us to ask respondents to consider a total of nine factors in terms of LNG regulation: local environmental impact, global environmental impact, security of natural gas supply, impact on prices of natural gas, geopolitics/foreign policy, free trade agreement with importing country, private property rights, impact on landowners and surrounding communities, and emergencies or disasters. All nine factors were considered important for regulation by our respondents, as the mean response for each factor was between "a moderate amount" and "a lot" in terms of the how the factor should be considered in regulation. Therefore, we computed a single index and used Cronbach's alpha to assess the reliability of the index, noted in Table 5.1 as $\alpha = 0.89$.

Benefit and risk perceptions. The measures of respondents' perceptions of benefits and risks associated with LNG exporting were determined from 10 survey items that asked respondents to indicate the level to which they believed each factor to benefit and the level they believed each factor to be at risk of being harmed by the United States exporting LNG to other countries, on a scale of "No benefit/risk at all" (1) to "A great deal of benefit/risk" (5). The 10 factors were US jobs, carbon emissions, US energy security, local environmental impact, US energy costs, you (the respondent) personally, US economy, public health and safety, private property rights, and hydraulic fracturing (fracking). We compared respondents' overall average perceived benefit and average perceived risk for

[1] MTurk response times: mean = 423 s; median = 321 s; minimum = 66 s; 95{sup|th} percentile = 72 s or longer.

Table 5.1 Reliability analysis for regulatory policy preference index.

Survey item	Mean[1]	Standard deviation	Item total correlation	Alpha (α) if item deleted
Local environmental impact	3.87	1.14	0.75	0.87
Impact on landowners and surrounding communities	3.82	1.11	0.74	0.87
Emergencies or disasters	3.82	1.11	0.72	0.87
Security of natural gas supply	3.82	1.08	0.67	0.88
Global environmental impact	3.77	1.22	0.69	0.88
Impact on prices of natural gas	3.66	1.13	0.57	0.89
Geopolitics/foreign policy	3.60	1.06	0.60	0.88
Private property rights	3.60	1.18	0.57	0.89
Free trade agreements	3.53	1.11	0.53	0.89
Policy preference index	3.72	0.83		0.89

[1]Items measured on a five-point scale: "Not at all" (1); "A little" (2); "A moderate amount" (3); "A lot" (4); and "A great deal" (5).

each of the 10 survey items. Four items—US jobs, US energy security, US energy costs, and US economy—had net positive differences, indicating these dimensions of LNG export were perceived more as benefits than risks. The other six factors all resulted in negative differences, indicating that they aligned more as risks. These results also align with previous analysis of perceived benefits and risks of fracking (Thomas et al., 2017). We then constructed aggregate scale measures of benefits ($\mu = 2.68$, $\alpha = 0.83$) and risks ($\mu = 2.77$, $\alpha = 0.87$). When conducting reliability analysis, the resulting Cronbach's alpha calculations indicated to exclude two risk items (risk to you personally and risk of fracking), resulting in a final four-item risk index.

Other independent variables. Table 5.2 presents the other independent variables incorporated in our analysis. Demographic variables included age, gender, race/ethnicity, income, and education level. To account for political ideology, we used a five-point scale from "very liberal" (1) to "very conservative" (5). Our measure of awareness of LNG export came from a single-item asking respondents to indicate how much they had ever heard or read about the United States exporting natural gas to other countries, on a scale of "not at all" (1) to "a great deal" (5). As shown in Table 5.2, overall familiarity was low, with almost half of the respondents

Table 5.2 Frequencies of relevant independent variables for analysis.

Variable	Question/categories	Frequency or mean
Age	Please indicate your age	M = 37.33; SD = 12.54
Gender	Please indicate your gender: Male/Female/Prefer not to answer	49.0% female
Race/ ethnicity	Please indicate your race: White, non-Hispanic; Hispanic; Black, non-Hispanic; Other non-Hispanic; Prefer not to answer	79.8% white
Income	Please indicate your annual household income: (1) Less than $25,000; (2) $25,000 to $49,999; (3) $50,000 to $74,999; (4) $75,000 to $99,999; (5) $100,000 to $124,999; (6) More than $125,000; Prefer not to answer	50.0% income $50,000 or more
Education	Please indicate the highest level of education you have attained. Less than high school; High school graduate; Some college/Associate degree/Technical degree; Bachelor's degree or higher; Prefer not to answer	51.4% bachelor's degree or more
Political ideology	In general, your political view is: (1) Very liberal; (2) Liberal; (3) Moderate; (4) Conservative; (5) Very Conservative	M = 2.72; SD = 1.12
Awareness	How much have you ever heard or read about the US exporting natural gas to other countries? Not at all; A little; A moderate amount; A lot; A great deal	53.5% aware a little or more
State	Please enter the state you are a resident of. Residing in a state with existing, approved, or proposed export terminal: AK, FL, GA, LA, MD, MS, OR, & TX	24.7% in state with existing, approved, or proposed export

indicating that they had no familiarity with LNG exporting. As a result, this survey item was dichotomized into 1 = aware a little or more, 0 = not at all aware. One last factor that we decided to control for was whether the respondent's state of residence had existing, approved, or proposed LNG export terminals. Residing in such a state would seem to increase the likelihood that a respondent has had some experience with the LNG industry, and previous research has found that industry activities correlated with higher support for fracking (Boudet et al., 2016). At the time of the survey, the following states had existing, approved, or proposed LNG export terminals: Louisiana, Alaska, Texas, Maryland, Georgia, Mississippi, Florida, and Oregon. A dummy variable was created such that 1 = respondents in these states, 0 = all other respondents.

Method. Our analysis uses ordinary least squares regression analysis to identify significant correlations with benefits, risks, and regulatory policy preferences. We constructed a total of six models. First, we regressed the benefit and risk indices on demographics, ideology, awareness, and state LNG status. Then, we examined the regulatory policy preference, first with the same model, then adding the benefit and risk indices as independent variables. This approach is done to better capture the nuance of benefit and risk perceptions of LNG export. In all six models, we have standardized all nondummy independent variables to facilitate comparison of magnitude. In order to account for heteroskedasticity, we used robust standard errors, clustered by respondent state of residence.[2] We computed variance inflation factors for each model, with the results indicating no evidence of multicollinearity.

Results

First, we examined which factors influenced a respondent's benefit and risk perceptions of LNG export (Table 5.3). In terms of benefit perceptions (Model A1), we found that respondents with incomes over $50,000 ($\beta = 0.12$, $P = .017$), at least a bachelor's degree ($\beta = 0.11$, $P = .046$), politically conservative ($\beta = 0.10$, $P = .003$), and at least a little aware of LNG export ($\beta = 0.25$, $P < .001$) had higher levels of perceived benefits from LNG export. Notably awareness of LNG export had the strongest association with benefit perception of all factors in the model. Female ($\beta = -0.23$, $P = .002$) and white ($\beta = -0.24$, $P = .003$) respondents had significantly lower perceptions of benefits, relative to male and nonwhite respondents, respectively. Additionally, respondents residing in states with existing, approved, or proposed LNG export terminals ($\beta = -0.17$,

[2] Additionally, we conducted alternatively specified models, including multilevel models with individuals and state as the two levels. The results of the multilevel modeling aligned closely with the OLS models presented.

Table 5.3 OLS regression standardized coefficients for LNG export benefit and risk indices.

IV	Model A1: benefit index	Model A2: risk index
Age	0.0009	−0.0349
	0.9775	*0.1926*
Female	−0.2297**	0.2833***
	0.0015	*<0.001*
White	−0.2394**	0.1921*
	0.0028	*0.0317*
Income >50k	0.1219*	−0.0045
	0.0174	*0.944*
Bachelor's degree	0.1119*	0.0723
	0.0459	*0.2344*
Political ideology	0.0970**	−0.3258***
(Liberal to conservative)	*0.0025*	*<0.001*
Aware at least a little	0.2509***	0.1295
	<0.001	*0.0511*
State LNG export status	−0.1667***	0.0642
	<0.001	*0.5672*
Constant	2.7689***	2.3674***
N	1011	1011
F	19.7214	23.2559
R^2	0.0662	0.1119
Adjusted R^2	0.0588	0.1048

$P < .001$) perceived lower levels of benefits than respondents in other states. Overall, the model accounted for approximately 7% of the variation in the benefit index.

For the risk perception index (Model A2), we found the mirror-image of benefit perceptions for gender, race/ethnicity, and political ideology. Female ($\beta = 0.28$, $P < .001$) and white ($\beta = 0.19$, $P = .032$) respondents perceived higher levels of risks for LNG export, relative to men and nonwhite respondents. Meanwhile more conservative ($\beta = -0.33$, $P < .001$) respondents perceived lower levels of risk than liberal respondents. The political ideology variable had the strongest association with risk perception of all predictors in the model. Although a significant predictor of benefit perception, awareness of LNG export ($\beta = 0.13$, $P = .051$) was not significantly associated with risk perception using a 95% confidence interval—however just barely. Though not statistically significant, it is important to note that the directionality remains the same as with benefits. Age, income, education, and residence in a state exposed to LNG export were not significantly associated with risk perception. Overall, the model accounted for 11% of the variation in the risk perception index.

The results of policy preference regressions are presented in Table 5.4. In the baseline model (Model B1), we found four factors to have significant correlations with regulatory policy preferences: age, gender, political ideology, and awareness of LNG export. Older respondents ($\beta = 0.07$, $P = .007$) favored higher degrees of regulation, as did female respondents relative to male ($\beta = 0.21$, $P < .001$). Respondents with more conservative political ideology ($\beta = -0.20$, $P < .001$) preferred lower regulation consideration, as did respondents who were at least a little familiar with LNG export ($\beta = -0.21$, $P < .001$). Race/ethnicity, income, education, and residence in a state with LNG terminal operations were found to have no significant association with policy preference. The baseline model explained approximately 10% of the variation in regulation policy preferences.

The final three models (Models B2–B4) incorporate the perceived benefits and risk indices to the baseline model. The four significant factors in the baseline model—age, gender, political ideology, and awareness of LNG export—remain significant across all models, though with different magnitudes. Further, the four insignificant factors in the baseline model similarly have no significant relationship when accounting for risk and benefit perceptions. In Model B2, the LNG export risks index ($\beta = 0.31$, $P < .001$) was the strongest predictor of policy preference, with respondents perceiving higher levels risks from LNG export favoring a higher degree of regulation. The inclusion of the risk index improved model performance, as Model B2 explained 22% of the variance in policy preference.

Model B3 included the baseline predictors plus the perceived benefits of LNG export index. Unlike risk perceptions, respondents' perceptions of LNG export benefits ($\beta = 0.03$, $P = .371$) were not significantly associated with different levels of regulatory policy preferences. Model B3 explains roughly the same proportion of variance as the baseline. Finally, Model B4 included both risk and benefit perception in the same model. Again, respondents' perceived risks of LNG export ($\beta = 0.31$, $P < .001$) were the strongest predictor of regulatory policy preference, even when accounting for perceived benefits. As with the previous models, older respondents

Table 5.4 Linear regression standardized coefficients for LNG export regulation preferences.

IV	Model B1: baseline	Model B2: baseline and risk	Model B3: baseline and benefit	Model B4: baseline, risk, and benefit
Age	0.0699**	0.0798***	0.0699**	0.0798***
	0.0073	*<0.001*	*0.0064*	*<0.001*
Female	0.2121***	0.1323**	0.2183***	0.1416***
	<0.001	*0.0012*	*<0.001*	*<0.001*
White	−0.083	−0.1371	−0.0765	−0.1272
	0.2142	*0.0602*	*0.2673*	*0.0896*
Income >50k	0.0166	0.0178	0.0132	0.0126
	0.7527	*0.6775*	*0.8018*	*0.7707*
Bachelor's degree	−0.0315	−0.0519	−0.0346	−0.0569
	0.6072	*0.3421*	*0.5773*	*0.3061*
Political ideology (Liberal to conservative)	−0.2048***	−0.1130***	−0.2074***	−0.1166***
	<0.001	*<0.001*	*<0.001*	*<0.001*
Aware at least a little	−0.1702***	−0.2067***	−0.1771***	−0.2178***
	<0.001	*<0.001*	*<0.001*	*<0.001*
State LNG export status	−0.0583	−0.0764	−0.0538	−0.0693
	0.3252	*0.0844*	*0.3728*	*0.1205*
LNG export risk index		0.3058***		0.3081***
		<0.001		*<0.001*
LNG export benefit index			0.0266	0.0423
			0.3721	*0.1463*
Constant	3.7964***	3.9102***	3.7940***	3.9071***
N	1011	1011	1011	1011
F	17.5159	96.6129	15.7951	89.7275
R^2	0.0968	0.2175	0.0978	0.2199
Adjusted R^2	0.0896	0.2105	0.0897	0.2121

A White's test found that heteroskedasticity was present in the data. In response, cluster robust standard errors have been used for all four models, clustered by State of respondent, in order to account for varying levels of LNG production and regulation from state to state. Variance inflation factors were computed for each independent variable in the models, finding no factor greater than 1.8, indicating minimal presence of multicollinearity. Ramsey RESET tests returned insignificant results, indicating that the model is likely not omitting a polynomial form of an included variable. * = $P < .05$; ** = $P < .01$; and *** = $P < .001$. All four model F-statistics were significant at $P < .001$ level.

($\beta = 0.08$, $P < .001$) and women ($\beta = 0.14$, $P < .001$) preferred higher levels of regulation, while conservative ($\beta = -0.12$, $P < .001$) and respondents familiar with LNG ($\beta = -0.22$, $P < .001$) preferred lower levels.

Discussion and policy implications

The results of our analysis offer some predictable and some surprising answers to our five research questions. In line with findings from tangential fields, age was positively associated with support for regulation in all four models, though the magnitude was the smallest of any significant factor across all four models. Gender also had a significantly positive association with support for regulation, again aligning with research from similar areas to LNG export regulation. In the baseline model, gender had the strongest association with policy preference ($\beta = 0.21$), but this association was weakened with the addition of the risk index (Models B2 and B4). This finding can be explained by the correlation between gender and risk perception (Model A2), as women were found to perceive higher levels of risk from LNG export than men. Political ideology also was significant across all four policy preference models, with conservative ideology correlated with lower desire for regulation. And just as with gender, the correlation was weakened by the inclusion of the risk index in Model A2. This means that an individual's political ideology influences their regulatory policy preferences both directly and indirectly through an apparent political polarization of risks.

Awareness of LNG export had a strong correlation with perceiving higher levels of benefits, while also having a weaker, borderline-significant association with higher levels of risk perception. This relationship bore out in terms of policy preference, as respondents at least a little aware of LNG export preferred lower levels of regulation. Kahan et al. (2009) noted in their research of nanotechnology that "[p]eople who have a pro-technology cultural orientation are thus more likely to become exposed to information about nanotechnology and to draw positive inferences from what they discover" (p. 89). Such a process could be present with LNG export, with those that are more aware placing more emphasis on the benefits than the risks. However, interpretation of this variable should consider that the measure being used is a single survey item, which means there is less ability to mitigate the possibility of measurement error.

Lastly, we found that while risk perceptions were correlated with regulation preference, benefit perceptions were not. Respondents who perceived higher level of risk had significantly higher preference for regulation of LNG export. However, benefit perception had no significant correlation with policy preference. Taken together, it would appear that perception of potential risk outweighs any perceived benefit of LNG export when it comes to regulation. Given that the United States was a net importer of natural gas and other fossil fuels for so long, it is possible that the notion of exporting fuel is seen as more of a risk given historic energy crises such as the 1973 OPEC oil embargo and the 1990 oil price shock following the Iraqi invasion of Kuwait. It is also possible that the benefits are overstated, as respondents in states exposed to LNG export perceived significantly lower benefits than respondents in other states. For policymakers contemplating regulation of LNG export, the implication is to prioritize public concerns over risks and resist the urge to deregulate to maximize benefits. However, the interplay of awareness and benefit perceptions also indicate that the potential gains of LNG export are not as apparent to unfamiliar members of the public. As US exports continue to grow, perhaps the public will become more familiar and shift the balance on regulatory preferences.

One final note, our analysis of benefit and risk perceptions resulted in two indices that align very closely with the dominant frames of supporters and opponents of LNG export (Trang et al., 2019). Supporters emphasize the economic and geopolitical benefits of export, while opponents consider the environmental, health, and safety risks. For regulators, it appears that this is the standard story that will have to be navigated in any debate over regulation of LNG export. These competing perspectives will be difficult to reframe, and therefore will continue to be the dominant planes upon which regulatory debate will take place. Regulators will have to strike the right balance between benefits that mostly accrue at the national level (e.g., to the US economy and energy security) and risks that tend to occur locally (e.g., local environmental impacts, private property rights).

References

Adams, J. S. (2014). US LNG exportation: The regulatory process and its practical implications. *Journal of World Energy Law and Business, 7*(6), 582–594. https://doi.org/10.1093/jwelb/jwu034

Ansolabehere, S., & Konisky, D. M. (2009). Public attitudes toward construction of new power plants. *Public Opinion Quarterly, 73*(3), 566–577. https://doi.org/10.1093/poq/nfp041

Ansolabehere, S., & Konisky, D. (2014). *Cheap and clean how Americans think about energy in the age of global warming*. MIT Press.

Berinsky, A. J., Huber, G. A., & Lenz, G. S. (2012). Evaluating online labor markets for experimental research: Amazon.com's mechanical turk. *Political Analysis, 20*(3), 351–368. https://doi.org/10.1093/pan/mpr057

Bernstein, P., Tuladhar, S. D., & Yuan, M. (2016). Economics of U.S. natural gas exports: Should regulators limit U.S. LNG exports? *Energy Economics, 60*, 427–437. https://doi.org/10.1016/j.eneco.2016.06.010

Boudet, H., Bugden, D., Zanocco, C., & Maibach, E. (2016). The effect of industry activities on public support for 'fracking'. *Environmental Politics, 25*(4), 593–612. https://doi.org/10.1080/09644016.2016.1153771

Boudet, H., Clarke, C., Bugden, D., Maibach, E., Roser-Renouf, C., & Leiserowitz, A. (2014). "Fracking" controversy and communication: Using national survey data to understand public perceptions of hydraulic fracturing. *Energy Policy, 65*, 57–67. https://doi.org/10.1016/j.enpol.2013.10.017

Bousky, J. F., & Harrison, J. L. (2016). Final destination: Issues of liability for exporters of U.S. LNG implicated by recent DOE/FE decisions. *Energy Law Journal, 37*(1).

Chanis, J. (2012). U.S. liquefied natural gas exports and America's foreign policy interests. *American Foreign Policy Interests, 34*(6), 329–334. https://doi.org/10.1080/10803920.2012.742409

Clarke, C. E., Budgen, D., Hart, P. S., Stedman, R. C., Jacquet, J. B., Evensen, D. T. N., & Boudet, H. S. (2016). How geographic distance and political ideology interact to influence public perception of unconventional oil/natural gas development. *Energy Policy, 97*, 301–309. https://doi.org/10.1016/j.enpol.2016.07.032

Davis, C. (2012). The politics of "fracking": Regulating natural gas drilling practices in Colorado and Texas. *Review of Policy Research, 29*(2), 177–191. https://doi.org/10.1111/j.1541-1338.2011.00547.x

Davis, C., & Fisk, J. M. (2014). Energy abundance or environmental worries? Analyzing public support for fracking in the United States. *Review of Policy Research, 31*(1), 1–16. https://doi.org/10.1111/ropr.12048

Delborne, J. A., Hasala, D., Wigner, A., & Kinchy, A. (2020). Dueling metaphors, fueling futures: "Bridge fuel" visions of coal and natural gas in the United States. *Energy Research & Social Science, 61*, 101350. https://doi.org/10.1016/j.erss.2019.101350

Dietz, T., Dan, A., & Shwom, R. (2007). Support for climate change policy: Social Psychological and social structural influences. *Rural Sociology, 72*(2), 185–214. https://doi.org/10.1526/003601107781170026

Evensen, D., & Stedman, R. (2017). Beliefs about impacts matter little for attitudes on shale gas development. *Energy Policy, 109*, 10–21. https://doi.org/10.1016/j.enpol.2017.06.053

FERC. (2020). *LNG maps exports*. Federal Energy Regulatory Commission. https://www.ferc.gov/media/lng-maps-exports.

Flynn, J., Slovic, P., & Mertz, C. K. (1994). Gender, race, and perception of environmental health risks. *Risk Analysis, 14*(6), 1101–1108. https://doi.org/10.1111/j.1539-6924.1994.tb00082.x

Gilbert, A. Q., & Sovacool, B. K. (2017). US liquefied natural gas (LNG) exports: Boom or bust for the global climate? *Energy, 141*, 1671–1680. https://doi.org/10.1016/j.energy.2017.11.098

Gravelle, T. B., & Lachapelle, E. (2015). Politics, proximity and the pipeline: Mapping public attitudes toward Keystone XL. *Energy Policy, 83*, 99–108. https://doi.org/10.1016/j.enpol.2015.04.004

Gronholt-Pedersen, J. (2016). *First U.S. shale gas exports imminent as tanker docks at Sabine Pass*. https://www.reuters.com/article/us-cheniere-energy-lng-idUSKCN0VV0SK.

Hazboun, S. O. (2019). A left coast 'thin green line'? Determinants of public attitudes toward fossil fuel export in the Northwestern United States. *Extractive Industries and Society, 6*(4), 1340–1349. https://doi.org/10.1016/j.exis.2019.10.009

Hazboun, S., & Boudet, H. S. (2020). Public preferences in a shifting energy future: Comparing public views of eight energy sources in North America's Pacific Northwest. *Energies, 13*(8), 1940. https://doi.org/10.3390/en13081940

Hazboun, S. O., & Boudet, H. S. (2021). Natural gas – friend or foe of the environment? Evaluating the framing contest over natural gas through a public opinion survey in the Pacific Northwest. *Null*, 1–14. https://doi.org/10.1080/23251042.2021.1904535

Kahan, D. M., Braman, D., Slovic, P., Gastil, J., & Cohen, G. (2009). Cultural cognition of the risks and benefits of nanotechnology. *Nature Nanotechnology, 4*(2), 87–90. https://doi.org/10.1038/nnano.2008.341

Klump, E. (2019a). *5 things the gas industry is watching*. https://www.eenews.net/energywire/stories/1061164131/search?keyword=LNG+export.

Klump, E. (2019b). *Trump's DOE just agreed with Obama on climate*. https://www.eenews.net/energywire/stories/1061141171.

Malin, S. A., Mayer, A., Shreeve, K., Olson-Hazboun, S. K., & Adgate, J. (2017). Free market ideology and deregulation in Colorado's oil fields: Evidence for triple movement activism? *Environmental Politics, 26*(3), 521–545. https://doi.org/10.1080/09644016.2017.1287627

McCright, A. M., Dentzman, K., Charters, M., & Dietz, T. (2013). The influence of political ideology on trust in science. *Environmental Research Letters, 8*(4). https://doi.org/10.1088/1748-9326/8/4/044029

Medlock, K. B., Jaffe, A. M., & O'Sullivan, M. (2014). The global gas market, LNG exports and the shifting US geopolitical presence. *Energy Strategy Reviews, 5*, 14–25. https://doi.org/10.1016/j.esr.2014.10.006

Pierce, J. J., Boudet, H., Zanocco, C., & Hillyard, M. (2018). Analyzing the factors that influence U.S. public support for exporting natural gas. *Energy Policy, 120*, 666–674. https://doi.org/10.1016/j.enpol.2018.05.066

Stedman, R. C., Evensen, D., O'Hara, S., & Humphrey, M. (2016). Comparing the relationship between knowledge and support for hydraulic fracturing between residents of the United States and the United Kingdom. *Energy Research and Social Science, 20*, 142–148. https://doi.org/10.1016/j.erss.2016.06.017

Thomas, M., Pidgeon, N., Evensen, D., Partridge, T., Hasell, A., Enders, C., Herr Harthorn, B., & Bradshaw, M. (2017). Public perceptions of hydraulic fracturing for shale gas and oil in the United States and Canada. *Wiley Interdisciplinary Reviews: Climate Change, 8*(3). https://doi.org/10.1002/wcc.450

Trang, T., Taylor, C. L., Boudet, H. S., Keith, B., & Peterson, H. L. (2019). Using concepts from the study of social movements to understand community response to liquefied natural gas development in Clatsop county, Oregon. *Case Studies in the Environment, 1*–7. https://doi.org/10.1525/cse.2018.001800

University of Texas Energy Poll. (2013). *UT Energy Poll shows consumer opposition to exporting natural gas*. https://news.utexas.edu/2013/04/09/spring-2013-ut-energy-poll-shows-consumer-opposition-to-exporting-natural-gas.

U.S. EIA. (2019a). *LNG export to Europe increase amid declining demand and spot LNG prices in Asia*. https://www.eia.gov/todayinenergy/detail.php?id=40213.

U.S. EIA. (2019b). *Natural gas explained: Liquified natural gas*. https://www.eia.gov/energyexplained/natural-gas/liquefied-natural-gas.php.

U.S. EIA. (2019c). *Natural gas weekly update*. https://www.eia.gov/naturalgas/weekly/.

U.S. EIA. (2021). *Asia became the main export destination for growing U.S. LNG exports in 2020—Today in Energy*. https://www.eia.gov/todayinenergy/detail.php?id=47136.

CHAPTER 6

Energy and export transitions: from oil exports to renewable energy goals in Aotearoa New Zealand

Patricia Widener
Sociology, Florida Atlantic University, Boca Raton, FL, United States

Island transitions

The island nation of Aotearoa New Zealand is little known as an oil exporter. Yet for decades, it has extracted relatively small amounts of oil for export and greater amounts of natural gas for domestic consumption. Beginning around 2010, however, growing global demand, coupled with new and unconventional technologies including deep water offshore drilling and onshore hydraulic fracturing, supported industry and political interests in expanding offshore and onshore exploration. With a history of exporting primary goods (including dairy, meat, wool, and timber), New Zealand was being promoted as a potential major exporter of oil. In response, pro- and antidrilling groups formed to debate the local and national rewards and pitfalls of large-scale extraction for export. Oil and gas corporations headquartered in Europe and North and South America, including Anadarko, Chevron, OMV, Petrobras, Royal Dutch Shell, and Statoil, had received permits to begin seismic testing and/or exploration in the country's frontier waters. That is until 2018, when future permits were banned.

This study began in 2013 to explore how New Zealanders, with a national identity and international reputation of being environmental stewards (Clements, 1988; King, 1986; Tucker, 2011), would respond to a political and economic agenda of becoming an oil and gas frontier in a time of climate change. At the time, political and economic leaders were projecting that the island nation could become the "Texas of the South Pacific" (see Loomis, 2017); and in running toward this status, the state projected a doubling of the country's oil production between 2010 and

Public Responses to Fossil Fuel Export
ISBN 978-0-12-824046-5
https://doi.org/10.1016/B978-0-12-824046-5.00005-9

2030 (Ministry of Economic Development, 2011, p. 3). Importantly, the timing of this announcement (and this study) coincided in the middle of three rounds of national parsing on energy production and climate change. Between 2000 and 2020, the nation witnessed a rise in climate awareness and political discourse, beginning with Labor Party Prime Minister Helen Clark (1999–2008), who appointed a climate change minister within the Ministry for the Environment and who had proposed becoming the world's first carbon neutral nation, while simultaneously expressing interest in expanding oil and gas block offers independent of climate change discussions. With a narrow vote, New Zealand ratified the legally binding the Kyoto Protocol in 2002 and proposed a carbon tax (which was rejected by a strong farming lobby). Clark was followed by John Key (2008–16) of the National Party, who withdrew from the Kyoto Protocol, while launching an eight-step Petroleum Action Plan to expand the country's commitment to oil exports. At the time, the Parliamentary Commission for the Environment (2014, p. 5) indicated that gas extraction had been identified as a "sunset industry" before the shift toward intensifying exploration for both oil and gas through deep water exploration and onshore hydraulic fracturing. During the proextraction period, the industry's perceived economic opportunities "overwhelmed opportunities to reduce emissions" (Murphy & Murphy, 2012, p. 265). In explaining the situation to me in 2013, one antidrilling activist said, "Both major parties are hell bent" on this energy transition *toward* extracting oil and natural gas.

Then in 2017, a political switchback to the Labor Party led to a climate-committed leader, Prime Minister Jacinda Ardern, who linked domestic climate action and proextraction policies, leading to a groundbreaking ban on future permits for offshore oil and gas exploration. Unlike preceding governments, the Ardern government connected the nation's climate goals to ending offshore exploration for export. Existing offshore permits and activities would be honored, but all new exploratory bids and permits were halted in 2018. After nearly a decade of mobilized resistance to offshore oil exploration and gas to a lesser extent (see Diprose et al., 2016; O'Brien, 2013; Widener, 2018a, 2021), antidrilling, climate, and marine justice activists had an ally in political leadership. During the peak of antidrilling activism, many climate activists had interpreted stopping oil and gas extraction as a significant global climate action (Widener, 2021), which was partially achieved with the offshore exploration ban and the signal the ban sent on closing the doors to widescale oil and gas activities. As some coastal communities celebrated, a press release by Greenpeace New Zealand quoted the group's executive director as claiming that "the tide has turned irreversibly against Big Oil in New Zealand" (Greenpeace New Zealand, 2018).

In response to the ban, oil and gas companies still operating in the country shifted their focus to reworking or expanding existing sites of extraction in Taranaki, the traditional region of extraction along the southwest coast of the North Island. Others left when seismic testing or offshore exploration failed to yield results of commercial value or quantity. Subsequently, the state altered its projections from doubling exports to a decline in oil production from approximately 11 million barrels of oil in 2019 to zero by 2039 (Ministry of Business, Innovation and Employment, 2019, p. 41). Of oil extracted in 2018, more than 90% was exported, while the majority of oil consumed was imported primarily from the United Arab Emirates (Minister of Business, Innovation and Employment, 2019, p. 45). In relative terms, the United States consumed more than 20 million barrels per day in 2019, according to the US Energy Information Administration; while New Zealand, with a population of close to five million, consumed approximately 50 million barrels per year (Ministry of Business, Innovation, and Employment, 2019, p. 46).Yet when thinking through the lens of social, climate, and environmental justice movements, the act of a small nation banning offshore exploration to protect their seas and to act globally on climate change was monumental. New Zealand elected not to participate in the global supply of petroleum, much like Belize, Costa Rica, Denmark, France, and Ireland, which have been identified as the "global first movers" toward implementing policies to "keep it in the ground" (Carter & McKenzie, 2020, p. 1344; see also; Gaulin & Le Billon, 2020). Environmental activists in other nations, including those mobilizing in defense of the Amazon, have tried, but were ultimately unsuccessful (Martin, 2011).

To analyze how coastal communities and environmental and climate activists responded to the nation's transition to large-scale export-oriented exploration in frontier waters, I draw upon three cases. These include: (1) two coastal towns on the South Island competing to become the terminal site for offshore activities; (2) national disputes over gasoline prices and employment in the oil and gas sector; and (3) green jobs, carbon neutrality goals, and export aspirations of renewable energy technologies.

Critical time and place methodologies

Guided by political economy and social movement perspectives and the methodological relevance of time and place, I studied community, climate, and environmental activism in Aotearoa New Zealand between 2013 and 2014 (see Widener, 2018a,b, 2021). The guiding question in the broadest of terms was how would New Zealanders respond to a political and economic agenda of becoming an oil and gas frontier in a time of anthropogenic climate change. Like Hess's (2018) study of New York, there were multiple energy-transition coalitions. Some emphasized climate actions; some advocated for changes to transportation; others mobilized in defense of the ocean or their communities; still others acted against off-shoring drilling or against onshore hydraulic fracking; and some advocated for local green jobs and a reduction of domestic greenhouse gas emissions. Each one was connected to or aware and supportive of the others, and were arranged along a continuum of being community-centered, national, or global in scope and goal.

Qualitative data were collected through approximately 60 interviews, primarily but not exclusively with opponents of oil and gas exploration; observations of approximately 40 events, including town halls, conferences, protests, rallies, and public meetings on oil and gas proposals and energy alternatives; and a range of online or published documents from supporters and resisters of oil and gas exploration. Updated accounts and statistics and a subset of the original data are presented in this chapter.

Regarding the importance of time and place (Tuck & McKenzie, 2015), this study occurred when oil and gas frontiers were being opened for exploration, when climate awareness was growing, and shortly after two oil spills, including the 2010 *Deepwater Horizon* disaster in the Gulf of Mexico and a cargo wreck and oil spill off the coast of New Zealand in 2011. At this time, the industry was gaining offshore access into regions and near communities with no history of oil and gas extraction. These communities were learning for the first time about the industry and its potential employment and ecological impacts, and I was able to attend public rallies and community and government-organized meetings on the industry's proposals. Up to this point, industry activities, including both onshore and offshore extraction and port facilities, had occurred in the single province of Taranaki on the southwest of the North Island. In relative terms, Taranaki is the New Zealand version of Alberta or Texas, but on a smaller scale. The frontier coastal communities analyzed in this chapter include those on the southeast of the South Island, including the towns of Dunedin and Invercargill. Coastal residents in the frontiers and environmental activists nationwide discovered and simultaneously mobilized against the possibility and potential future of offshore oil and gas extraction in frontier waters and near frontier coastlines, galvanizing a national and ontological inquiry on whether or not to become a potentially major oil and gas supplier.

Debating offshore exploration and an onshore terminal

If oil and/or gas were found in export volumes off the east coast of the South Island, then onshore supply facilities would need to be built on a scale that could impact generations and alter the coastline for decades to come. While others have studied communities directly affected by a drilling project, few have examined community discovery and conflicts over the siting of supply or infrastructural projects (see Boudet et al., 2018; Tran et al., 2019). Yet, the industry is one with a wide socioecological and socioeconomic footprint, including multiple zones of extraction, land and sea transportation, storage and support facilities, refining, processing, and consumption. So as new bids and exploration rights were initiated off the South Island, two groups formed to advocate for and against exploration, while the business interests of two coastal communities offered competing arguments for the requisite onshore support terminal.

Even before oil and/or gas were found, industry proposals fueled tensions within the southeastern region and between two coastal towns, Dunedin and Invercargill. Few in the frontiers possessed any experience in oil and gas extraction, thus sharing a disadvantage in trying to understand for the first time a highly technical industry, whose national regulations and consents were less than transparent or were still being formulated by the appropriate state agencies.

Within the southeast province of Otago, two competing groups organized: Oil Free Otago and Progas Otago. As their names suggest, no one knew what would be found offshore in commercial supply: oil for export and/or natural gas for domestic consumption and/or export. Antidrilling activists argued that the progas contingent was hoping to find oil, which is of higher commercial value, but was claiming that their prodrilling advocacy supported natural gas as "bridge fuel," a product that also lacked the associated notoriety of an offshore oil spill.

One antidrilling resident indicated that everyone in the "small community [had] all kinds of friends or relatives or associates in the other [prodrilling] group." Another antidrilling resident emphasized the division as a hostility directed at them for just trying to offer alternatives for the future: "The anger and hostility and irrationality of the proindustry lobby is predictable, but that doesn't make it any less frustrating to deal with. We go to great lengths to try and present opinions grounded in science and evidence, and you get a whole lot of rhetoric [from them]." Another antidrilling activist identified the progas camp as one that had "a lot of money" and "sprung up incredibly quickly. It certainly doesn't feel like an organic group." These comments speak to a history of astroturf groups, rather than community-started grassroots organizations, entering a community to cultivate proindustry arguments and sentiments. Others believed that people in the region wanted to be perceived nationally and internationally as positive and pro-progress however defined. From this perspective, "there are many people who don't like to be perceived as negative, or antiprogress, or anti-anything," and who believe any employment should be courted.

In arguing to expand offshore exploration, the prodrilling contingent, including chambers of commerce, acted as a promotional arm of the industry to attract land and port infrastructure, activities, and employment. In the *Otago Daily Times*, a well-regarded and frequently referenced newspaper, the commerce chambers of Dunedin and Invercargill, competed in listing the top 10 reasons for the global industry to build along their coastlines. Keeping in mind that neither town had any direct experience with large-scale offshore extraction, Invercargill, the more industrial of the two, conveyed a technical and industrial knowledge of the priorities, scale, labor, and needs of offshore drilling. In contrast, Dunedin, a university town, promoted its hospital, university, hotels, and natural beauty to reflect its more educated workforce and lifestyle amenities. More cynically, the South Island's inexperience and limited working knowledge of the industry may have appealed to corporations wanting to write their own regulations and govern inexperienced local politicians and businesses.

Dunedin council members also weighed in, offering opposing views that aligned with the two divergent activist groups. Councilor Andrew Whiley took the position that "like it or not" the decision to explore had been made, so "let's embrace the opportunity to play host as the southern exploration hub for the companies that are coming" (Whiley, 2014, p. 9). Councilor Jinty MacTavish (2014, p. 9) countered: "My preference would be for our council, our people and our economy to be invested in resilient, future-focused industries that will continue to serve our community and the globe regardless of our climate future—things such as sustainable food production, renewable energy and education." Two days later, Dunedin Mayor Dave Cull meditated the two arguments. While acknowledging that if the focus was on climate change, many would argue a resounding "no" on drilling, but drilling was not a local decision: "[T]he decision on whether to drill, is not in this community's hands and the drilling, being off the coast, is outside the jurisdictional area of any local authority" (Cull, 2014, p. 15). Therefore, according to Cull, if drilling were going to happen off their shores with or without residents' consent, should Dunedin capture the purported support jobs and monetary benefits, and then use those funds to transition to a low-carbon future and sustainable employment in other areas? On the point of jurisdiction, exploration was permitted in national waters, rather than nearshore local waters.

One concern was that when primary and support industries and infrastructures settle in, like they had in Taranaki (and Aberdeen, Alberta, and Texas), community criticism recedes and livelihoods become intertwined in perception and reality with the industry's successes in such a way that residents eventually accept the risks and downplay the hardships (Freudenburg & Gramling, 1994). When thinking of site fights, areas already industrialized are more acquiescent to expansions than areas without previous industrial activities (see McAdam & Boudet, 2012; Widener, 2011). Fortunately for the antidrilling faction and unfortunately for the prodrilling camp, an onshore hub to support offshore exploration on the South Island never materialized once the ban was announced.

Disputing employment promotions and lowered petrol prices

Beyond offshore drilling and regional siting disputes, national responses were also mixed regarding the desire for employment and lowered gasoline prices. It is not that anyone advocated for higher prices or less work, but that the promotional narrative was challenged by more critical and comparative arguments. For context, in late 2020, New Zealanders paid approximately NZ$7.87 (US$5.43) per gallon of gasoline or more than double what US consumers were paying (approximately US$2.47 per gallon) (see www.GlobalPetrolPrices.com for weekly comparisons). Yet given the global supply and transport of oil and the country's limited capacity to refine oil for local consumption, antidrilling activists argued that even if large quantities were found offshore, the price at the pump would not change. In all likelihood, oil extracted in New Zealand would enter the global supply for international markets rather than domestic ones, which sounded counterintuitive to many residents who wanted lower prices. At the time, New Zealand exported crude oil and imported both crude and refined oil for consumption. With only one refinery, New Zealand lacked the capacity to refine its domestic supply. Likewise, the country remained committed to the extraction and consumption of its natural gas.

For a confused audience who just wanted lowered prices or independence from international suppliers, a Greenpeace organizer stressed that "none of this oil is for New Zealand. ...It belongs to the foreign company that drills it. ...They will transship it. It will never land in this country. It will never affect the price of oil at the pump in this country." In an interview, another activist told me, "We usually say it doesn't make any difference. ...The oil that's drilled here doesn't necessarily even enter into our market. We are not going to get our petrol just off the coast. We are still going to get it from Saudi Arabia or wherever." According to another antidrilling activist, "You could find Saudi Arabia here, but we'd still be paying the same at the petrol pump." The comparative import of Saudi Arabia was a reoccurring theme for promoters and resisters, in terms of potential scale, wealth, dependency, or decline in transparency or democratic practices.

The second point of tension centered on employment, including the potential for well-paid jobs on the offshore platforms or land-based support facilities, which ranged from physically demanding to science, engineering, and technology jobs. In the midst of local and national disputes on the pros and cons of offshore exploration and potential employment, antidrilling activists and community advocates walked a tightrope. "People say how can you pretend to represent this community and not want oil and gas turning up and the jobs that come with that?" one said. "For people who live on the dole, how can you not [assist] them by welcoming this job opportunity?" Similarly, another also recognized the dilemma:

> *People are terrified of the loss of their jobs and what that might mean to their livelihood and to their families, and I guess for those guys the oil and gas offer is, or what they perceive to be what oil and gas offers, is jobs. ...I have real questions about the resilience of that model and how sustainable it is. I think if we are serious about wanting to create jobs, we need to create them ourselves and within our communities and not rely on silver bullet multinational solutions that [are] made well outside of our city. But there's a deep belief in some parts of our community this is an answer. It will bring hundreds of jobs.*

From the recollection of another, some residents "will say we are just on this high level—or we can afford to be at this debate when they are just worried about their check or the next bill or their mortgage—and that becomes a difficult argument. ...If they talk about jobs, it's a bit tricky even though we do explain that there are far more jobs in the green sector or primary production sector than fossil fuels." Another referenced how prodrilling advocates refer to export-oriented employment as "wealth creating businesses," especially for smaller towns. To counter this argument, they promoted local jobs and greater energy efficiency. "You can pour as much 'wealth creating businesses' as you want into a community, but unless you are working at the other end to reduce the amount that is flowing out, that's just mad," one said. "You are not going to improve your outcomes."

Antidrilling campaigners believed that people would transition to clean, green, and environmentally sustainable jobs that avoid or amend environmental damage, or what is known as a key part of the just transition movement (Sze, 2020), if those jobs were available and well-paid and if the transition was "easy." To the argument "we have families to feed," one antidrilling activist suggested that people would be willing to embrace the idea of a local and green economy, "if it was there and if it was perceived as easy and lucrative."

The bind was that the "big, green, economy carrot" that one activist envisioned still needed to be planted, grown, and harvested. The oil and gas industry already had a global reputation of high-paying jobs across the labor-to-professional employment spectrum. And it possessed a misleading reputation for an abundance of them for local residents, even though the industry only employed about 11,000 people, according to the Petroleum Exploration and Production Association of New Zealand (PEPANZ, 2021). For comparison, tourism employed approximately 230,000 people (Tourism New Zealand, 2020) and the agricultural industry employed more than 86,000 (Statista, 2021). As it were, the oil and gas industry appeared to be rushing in and self-promoting its employment opportunities before green, blue-green, and just transition jobs were available, well known, and well regarded. The progreen camp knew their battle was uphill, which led one to acknowledge, "If there is any perceived loss or pain through the letting go of the oil and gas industry before the green economy was there to fill that void, then there would be a lot more kicking and screaming along the way."

Transitioning to green jobs and green technology exports

Climate, environmental, just transition, and antidrilling activists tried to counter arguments on employment, while promoting investments in domestic green jobs and the export of renewable energy technology. At the time, New Zealand was perceived by antidrilling activists as having the ability to shift toward clean technologies and renewable energy and toward becoming a global player on what remains an inevitable future. Indeed, the antidrilling campaigns reinvigorated climate activists, who had been silenced by climate deniers (Widener, 2021). At the time of this study, the country had already demonstrated its expertise in renewables by achieving 75% of its electricity through a mix of renewables, including geothermal, wind power, hydropower, and solar. Greenpeace New Zealand (2013), which commissions scientists and researchers to demonstrate and legitimate their arguments, suggested that the country could have 100% renewable energy by 2025; and the Ministry of Economic Development (2011, p. 6) offered an equally impressive commitment to 90% for the same year. Greenpeace also projected oil-free road transport by 2035 and world leadership in ocean energy technology. According to the group's report, New Zealand possessed at least 60 potentially world-class companies that could export clean energy technologies or climate solutions, including wood-based biofuels or ocean wave and geothermal energy. In an interview, one activist said, "New Zealanders find it hard to believe that we actually lead the world in many things, but you go around the world and you can see that we are absolutely the experts at geothermal."

For another renewable energy advocate: "I am still holding hope that New Zealand can be a leader in terms of showing the way to switch to other forms of energy, ...and that the quest for fossil fuels is purely for profit to sell overseas. So, if there is a single country on the face of the earth that could actually show the way with a small population and a highly renewable energy program already in place, it's this one." Still others believed that the renewable energy efforts were being stalled by ministers, who wanted to protect oil and gas exports. "We've heard from the renewable energy industry that when they meet with the ministers and say how New Zealand can go to 100% renewable electricity, the ministers say, that's not a good thing. We need the oil industry, the gas industry. [The industries] need to know that they can keep supplying gas and make some money locally because otherwise the costs of export are too great." This comment reflected the industry's political influence beyond their own operations.

The desire to lead where it could in renewable energy and sustainable technology as a counter to expanding fossil fuel exploration was present in interviews, informal conversations, public displays, and throughout the media. As climate activist suggested, "New Zealand can't be a global leader [in large-scale oil and gas extraction], but it could be a global solutions leader." This argument resonated in the hearts and minds of many local, clean energy advocates who wanted to be and to be seen as climate-informed and global environmental stewards. In contrast to exporting oil and building another support terminal, New Zealand could provide 100% renewable electricity domestically and export renewable energy knowledge, inspiration, and technology.

Even though antidrilling and climate activists were mobilizing for a rapid transition away from fossil fuels in terms of domestic consumption and global supply chains, scholars have indicated that a more likely scenario would involve "energy additions," rather than transitions (York & Bell, 2019, p. 40). After studying historic energy trends, York and Bell reasoned that both fossil fuel and renewable energy production would occur simultaneously for a time without necessarily a reduction in the former. Nations could also continue to export oil and gas, while becoming less reliant on fossil fuels domestically. Within the Organization for Economic Co-operation and Development nations, New Zealand has the third highest levels of electricity production through renewables following only Norway (a major offshore producer) and Iceland (an Arctic nation with the potential for offshore exploration) (Ministry of Business, Innovation, and Employment, 2019, p. 27). Those I interviewed saw the country as sitting on the edge of achieving international recognition for a low carbon, locally driven economy, before tumbling backward by the political drive to expand oil and gas exploration for export. On the consumption-side where greenhouse gas emissions are counted, New Zealand was well situated to increase its renewable energy sources and reduce its fossil fuel dependency. On the production side, however, the country was positioning itself to supply other nations the drug or bullet (or an addictive or threatening product), without legal responsibility for the consequences of exporting oil and facilitating the greenhouse gas emissions of others. But instead of accepting this turn, activists and residents mobilized and within a decade had installed more climate-engaged leaders, who reconciled the disconnect between the country's domestic climate policies and its export-oriented fossil fuel activities by banning offshore exploration as a climate action (and by beginning to address the greenhouse gas emissions by the country's economically significant and export-oriented livestock industries).

Conclusions

After the Labor Party gained seats in 2020, Ardern appeared to advance the country's climate commitments and global leadership once again. In late

2020, Ardern declared a "climate change emergency" and committed the government to carbon neutrality by 2025 and the country to carbon-neutrality by 2050. New Zealand is a nation to watch for its just transition and socioecological activism and for its current leadership and public stewardship on climate change and energy production and consumption.

In brief, critical oil and antidrilling narratives matter. Community actions and social movements build broad awareness, strengthen regulations, increase oversight, insert community involvement in key decisions, generate national discussions, and achieve bans or moratoria (Buttel, 2003; Ladd, 2018; McAdam & Boudet, 2012; Pellow, 2014; Widener, 2011, 2021). At one time, the state courted exploration in frontier waters as an appropriate export-to-riches economic model regardless of climate change. Yet, coastal communities and environmental and climate activists responded persuasively that there were alternatives; and by doing so, they may serve as aspiration for other global activists thwarted by climate inaction or proextraction lobbyists.

References

Boudet, H., Gaustad, B., & Tran, T. (2018). Public participation and protest in the siting of liquefied natural gas terminals in Oregon. In A. E. Ladd (Ed.), *Fractured communities: Risk, impacts, and protest against hydraulic fracking in US shale regions* (pp. 248–270). New Brunswick, New Jersey: Rutgers University Press.

Buttel, F. H. (2003). Environmental sociology and the explanation of environmental reform. *Organization and Environment, 16*(3), 306–344.

Carter, A. V., & McKenzie, J. (2020). Amplifying 'keep it in the ground' first-movers: Toward a comparative Framework. *Society and Natural Resources, 33*(11), 1339–1358.

Clements, K. (1988). *Back from the brink: The creation of a nuclear-free New Zealand.* Wellington: Allen & Unwin/Port Nicholson Press.

Cull, D. (January 22, 2014). *Opportunity, challenge in gas exploration.* Otago Daily Times.

Diprose, G., Thomas, A. C., & Bond, S. (2016). 'It's who we are': Eco-nationalism and place in contesting deep-sea oil in Aotearoa New Zealand. *Kōtuitui: New Zealand Journal of Social Sciences Online, 11*(2), 159–173.

Freudenburg, W. R., & Gramling, R. (1994). *Oil in troubled water: Perceptions, Politics and the battle over offshore drilling.* Albany, New York: State University of New York Press.

Gaulin, N., & Le Billon, P. (2020). Climate change and fossil fuel production cuts: Assessing global supply-side constraints and policy implications. *Climate Policy, 20*(8), 888–901.

Greenpeace New Zealand. (2013). *The future is here: New jobs, new prosperity and a new clean economy.* Auckland: Greenpeace New Zealand.

Greenpeace New Zealand. (April 12, 2018). *Ardern makes oil history: 'Huge win for climate and people power' — Greenpeace.* Press release. Retrieved from https://www.greenpeace.org/new-zealand/press-release/ardern-makes-oil-history-huge-win-for-climate-and-people-power-greenpeace/.

Hess, D. J. (2018). Energy democracy and social movements: A multi-coalition perspective on the politics of sustainability transitions. *Energy Research and Social Sciences, 40*(1), 177–189.

King, M. (1986). *Death of the Rainbow warrior*. Auckland: Penguin Books.
Ladd, A. E. (Ed.). (2018). *Fractured communities: Risk, impacts, and protest against hydraulic fracking in U.S. Shale regions*. New Brunswick, New Jersey: Rutgers University Press.
Loomis, T. M. (2017). *Petroleum development and environmental conflict in Aotearoa New Zealand: Texas of the South Pacific*. Lanham, Maryland: Lexington Books.
MacTavish, J. (January 20, 2014). *Fossil fuel position based on science, best interests*. Otago Daily Times.
Martin, P. L. (2011). *Oil in the soil: The Politics of paying to preserve the Amazon*. Lanham, MD: Rowman & Littlefield.
McAdam, D., & Boudet, H. S. (2012). *Putting social movements in their place*. Cambridge: Cambridge University Press.
Ministry of Business, Innovation and Employment. (2019). *Energy in New Zealand 19*. Wellington: Ministry of Business, Innovation and Employment. Retrieved from http://www.energymix.co.nz/our-consumption/new-zealands-consumption/. (Accessed 28 November 2020).
Ministry of Economic Development. (2011). *New Zealand energy strategy 2011-2021*. Wellington: Ministry of Economy Development. Retrieved from http://www.med.govt.nz/sectors-industries/energy/pdf-docs-library/energy-strategies/nz-energy-strategy-lr.pdf. (Accessed 22 January 2012).
Murphy, R., & Murphy, M. (2012). The tragedy of the atmospheric commons: Discounting future costs and risks in pursuit of immediate fossil-fuel benefits. *Canadian Review of Sociology, 49*(3), 247–270.
O'Brien, T. (2013). Fires and flotillas: Opposition to offshore oil exploration in New Zealand. *Social Movement Studies, 12*(2), 221–226.
Parliamentary Commission for the Environment. (2014). *Drilling for oil and gas in New Zealand: Environmental oversight and regulation*. Wellington: New Zealand Parliamentary Commission for the Environment.
Pellow, D. N. (2014). *Total liberation: The power and promise of animal rights and the radical earth movement*. Minneapolis: University of Minnesota Press.
PEPANZ. (2021). *The importance of oil and gas to New Zealand*. Retrieved from https://www.pepanz.com/oil-and-gas-new-zealand/the-importance-of-oil-and-gas-to-the-new-zealand-economy/. (Accessed 27 February 2021).
Statista. (2021). *Number of Employees in the agriculture industry in New Zealand from 2011 to 2020*. Retrieved from https://www.statista.com/statistics/1013227/new-zealand-employee-count-in-agriculture-industry/. (Accessed 28 February 2021).
Sze, J. (2020). *Environmental justice in a moment of danger*. Oakland: University of California Press.
Tourism New Zealand. (May 27, 2020). *About the tourism industry*. Retrieved from https://www.tourismnewzealand.com/about/about-the-tourism-industry/. (Accessed 28 February 2020).
Tran, T., Taylor, C. L., Boudet, H. S., Baker, K., & Peterson, H. (2019). Using concepts from the study of social movements to understand community response to liquefied natural gas development in Clatsop county, Oregon. *Case Studies in the Environment, 3*(1), 1–7.
Tucker, C. (2011). The social construction of clean and green in the genetic engineering resistance movement of New Zealand. *New Zealand Sociology, 26*(1), 110–121.
Tuck, E., & McKenzie, M. (2015). Relational validity and the "where" of inquiry: Place and land in qualitative research. *Qualitative Inquiry, 21*(7), 633–638.
Whiley, A. (January 20, 2014). *Gas hub gains a case of us or them*. Otago Daily Times.
Widener, P. (2011). *Oil injustice: Resisting and conceding a pipeline in Ecuador*. Lanham, Maryland: Rowman & Littlefield.

Widener, Patricia (2018a). Coastal people dispute offshore oil exploration: Toward a study of embedded seascapes, submersible knowledge, sacrifice, and marine justice. *Environmental Sociology, 4*(4), 405–418.

Widener, P. (2018b). National discovery and citizen experts in Aotearoa New Zealand: Local and global narratives of hydraulic fracturing. *Extractive Industries and Society, 5*(4), 515–523.

Widener, P. (2021). *Toxic and intoxicating oil: Discovery, resistance, and justice in Aotearoa New Zealand.* New Brunswick, New Jersey: Rutgers University Press.

York, R., & Shannon, E. B. (2019). Energy transitions or additions? Why a transition from fossil fuels requires more than the growth of renewable energy. *Energy Research and Social Sciences, 51*(1), 40–43.

CHAPTER 7

Trends in Norwegian views on oil and gas export

Gisle Andersen[1,3], Åsta Dyrnes Nordø[1] and
Endre Meyer Tvinnereim[2,3]
[1]Norwegian Research Centre (NORCE), Social Science Department, Bergen, Norway; [2]University of Bergen, Department of Administration and Organization Theory, Bergen, Norway; [3]University of Bergen, Centre for Climate and Energy Transformation (CET), Bergen, Norway

Background—Norway's role as an oil and gas exporter

Despite Norway's small population size, its production of crude oil and natural gas covers about 2% and 3% of global demand, respectively (Fæhn et al., 2017; Norwegian Petroleum, 2021). Yet, its electricity production is almost entirely based on renewable sources: hydropower (90%) and windfarms (8%) (Energy Facts Norway, 2021). This means that Norway exports almost all the oil and natural gas it produces, making Norway the third largest exporter of natural gas in the world, and the seventh largest exporter of oil (Fæhn et al., 2017; Norwegian Petroleum, 2021). Europe is the most important market, and natural gas from Norway covers 20%–25% of Europe's gas demand. Oil and gas constitute about half of the value of total Norwegian exports (47.5% in 2019) and the sector employs about 5% of the work force (von Brasch et al., 2019).

To understand the structure of the Norwegian debate on oil and gas exports, the placement and lack of visibility of physical infrastructure is crucial. Production is entirely situated offshore, and, with few exceptions, the production and export facilities (platforms, tankers, and pipelines) are not visible from shore due to distance or because its situated underwater. Eighty-five percent of the petroleum products are exported directly by pipelines (gas) on the seabed or loaded directly on to tankers (buoy-loading) far from the shore (Fæhn et al., 2017; Norwegian Petroleum, 2021). The remaining 15% is transported by tankers or pipelines to onshore refineries in Norway, before being exported by tankers. Onshore infrastructure is concentrated at a few major facilities in sparsely populated areas along the west coast (see Fig. 7.1). Export infrastructure is thus invisible to most of the Norwegian populace. Although extraction policies are increasingly debated

Public Responses to Fossil Fuel Export
ISBN 978-0-12-824046-5
https://doi.org/10.1016/B978-0-12-824046-5.00010-2

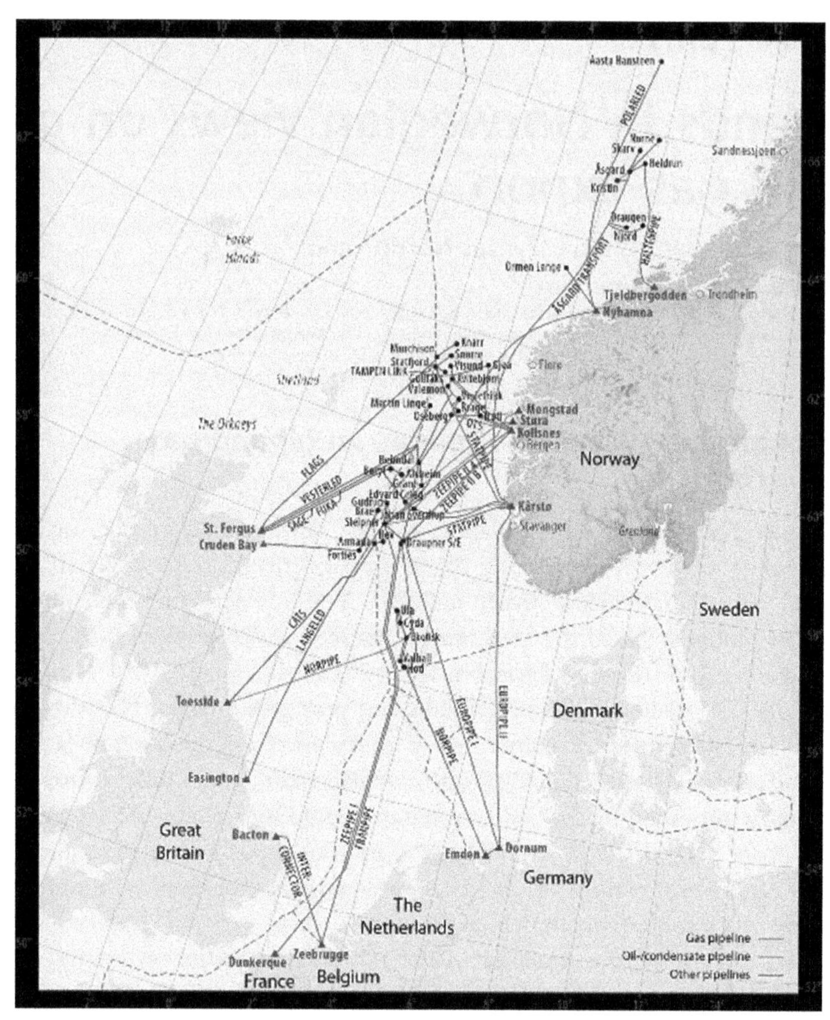

Figure 7.1 Map of Norwegian oil and gas export infrastructure. Updated March 2020. *(From Norwegian Petroleum Directorate (2020). https://www.norskpetroleum.no/en/?attachment_id=20687.)*

and contested because of greenhouse gas (GHG) emissions, the export infrastructure itself has not caused political conflicts or public protests. Moreover, there have been very few large oil spills from the production and export facilities, the spills have never reached shore and have received limited public attention (Andersen, 2017).

In addition to a nearly invisible export infrastructure, public debate and opinion on petroleum activities are shaped by successful redistribution of revenues from the sector to the population. The "Norwegian model" (Al-Kasim, 2006) gives democratic control over petroleum resources through a licensing system, and a special tax system gives the state a significant part of the production revenues from the sector (Ryggvik, 2018). Production and export increased rapidly from late 1980s, as did state income from the sector. The risk is that this creates a "Dutch disease" in the economy, leading to inflation and overspending in the public sector (Ramírez-Cendrero & Wirth, 2016). To avoid this the Norwegian sovereign wealth fund was set up in 1990. State income from the sector is invested in this fund, and only a share of the interests from the fund are used to fund public spending. The value of the fund increased rapidly from mid-1990s onward, and the current value is close to 11,000 billion NOK (ca. 1100 billion USD) (NBIM, 2021). The Norwegian petroleum regime also consists of incentives for further exploration drilling on the Norwegian Continental shelf. Most important, the Norwegian government carries a large part of the economic risk from exploration because 78% of the investment costs can be deducted from oil and gas company taxes.

Data and methods

How does the duality of being both one of the biggest oil and gas exporters in Europe and the invisibility of the export activities affect public policy and the public debate in Norway? We approach this question in two distinct ways. First, we draw on previous studies of the Norwegian political debate and discourse on fossil fuel and climate policies done by the authors (Andersen, 2017; Dale & Andersen, 2018) and others (Anker, 2018; Asdal, 2014; Bang & Lahn, 2019). Second, we analyze public opinion data from the Norwegian Citizen Panel (NCP) on issues related to oil and gas production and climate change, see, for example, Tvinnereim and Ivarsflaten (2016).

Policy analysis

Our policy analysis is based on previous published studies of Norwegian oil and gas and environmental policies (Andersen, 2017; Anker, 2018; Bang & Lahn, 2019; Dale & Andersen, 2018). None of these have analyzed oil and gas *export* exclusively, so we present the main points from these studies and discuss the consequences for the policy development on oil and gas export in light of the most recent policy developments in Norway. Andersen (2017) studies the development of environmental policies in Norway from

1945 to 2013, with an in-depth study of oil and gas policy as a main case. This analysis is based on verbatim reports from the Norwegian Parliament, white papers, propositions, and committee recommendations to Parliament and seeks to understand how decisions and policy positions are justified, and how this has changed over time. Dale and Andersen (2018) analyze the debate on the future of the oil and gas sector the last decade, with a particular focus on the role of energy scenarios. Anker (2018) summarizes much of the research by Norwegian historians on how prominent Norwegian politicians worked to reconcile a continued expansionist oil and gas policy with ambitious emissions reduction goals. Lahn and Bang (2019) combine a study of policy documents, media statements, and interviews with stakeholders and analyze the formation of new oil and gas policy coalitions after 2013.

Public opinion studies

The NCP fields online, probability-sample surveys among the Norwegian population two or three times a year. The panel was fielded for the first time in the fall of 2013. We analyze trend data from the survey descriptively by reporting mean results for relevant variables between 2013 and 2020. The variables we analyze in this chapter relate to public opinion in Norway on the expansion of petroleum drilling and on the future of the oil and gas industry. We choose these variables as these relate relatively closely to the issue of Norway's oil and gas exports. In addition, we analyze a background question related to respondents' worry about climate change in general.[1]

The first question asks whether the respondent supports or opposes oil and gas drilling in a vulnerable coastal area near the Arctic archipelagos of Lofoten and Vesterålen, and island of Senja. The areas are important for fisheries and host a growing tourist industry. The question is worded as follows:

"Consider the statements below. To what extent do you agree or disagree with them: … We should not allow oil and gas extraction in the area around Lofoten, Vesterålen and Senja." The response scale has seven points with the following values: *strongly agree—agree—somewhat agree—neither agree nor*

[1] Question codes from NCP are w02_km223abc_2, w03_r3dvh_1, r4dvh_1, r5dvh_1, r6dvh_1, r8km4_b, r8km4_b_panel, r8dvh_1, r10km4 r11pkkm4 r14pkkm4 (Lofoten oil/gas drilling); r6km66, r15km66 (development of the oil industry); w01_km36, w03_r3km23, r4km1, r5km6, r6km236, r7km2, r8bekym, r10km2, r11pkkm2_1, r14pkkm2, r17pkkm2 (worry about climate change).

disagree—somewhat disagree—disagree—strongly disagree. To analyze this question we make use of ten waves from 2013 until 2020.

The second item asks directly about how large the respondent thinks the oil and gas industry in Norway should be in 20–30 years. The vignette for this question reminds the respondents of the political debates on climate change and their links to the question of whether we should allow the industry to expand into new areas:

Production of oil and gas is an important industry for Norway. The industry provides many jobs and considerable income for society. At the same time, burning fossil energy is the main cause of climate change and activities on the Norwegian continental shelf contribute about 1/4 of Norway's greenhouse gas emissions. Opening new areas is also controversial because it can affect life in the sea and at the ice edge. The political parties have different opinions about how the petroleum industry should be developed over the next 20 to 30 years. There is disagreement over both how large the industry should be and whether new areas should be opened.

The subsequent question is worded as follows: "*Which of the following alternatives is closest to your view?*"

The response options are based on various points of view found in the public debate, and notably on party manifestos. They are worded as follows:

1. *In 20 to 30 years, the industry should be as big as possible. We should open up new exploration areas and, if possible, facilitate further growth.*
2. *In 20 to 30 years, the industry should be as big as it is today. We should take environmental precautions like we do today, open up new search areas where environmentally justifiable, and stimulate the maintenance of the industry.*
3. *In 20 to 30 years, the industry should be smaller than today. We should therefore not open up new areas, and exploration should be limited to areas where activity has already been established.*
4. *In 20 to 30 years, the industry should be considerably smaller than today. We should therefore not allow searches for new resources, and we should only permit production in existing fields until they have been depleted.*
5. *In 20 to 30 years, the industry should be discontinued. We should therefore not allow exploration for new fields. The production in existing fields should be considerably reduced within the next 5 years and closed down within the course of the time period.*

To analyze this question about the future of the oil and gas industry, we compare the two time points when it was asked first in spring 2016 (Wave 6) and then again in spring 2019 (Wave 15). The background question on worry about climate change is worded as follows: "*How worried are you about climate change?*"

Response options are given on a five-point scale, ranging from *not at all worried* to *very worried*.

The question probing worry about climate change is a standard question that has been repeatedly asked in the NCP, leaving us with 11 data points from 2013 until 2020.

Main findings from policy analysis

A number of policy studies have shown that by 1990 climate change had been accepted as real and as a severe societal problem by the majority of political parties in Norway (Andersen, 2017; Anker, 2018; Asdal, 2014; Nilsen, 2001). This sparked a number of policy processes to reduce GHG emissions. However, a majority of the political parties continued to support a further expansion and intensification of oil and gas extraction. Andersen (2017, pp. 370–405) shows that this decoupling of climate and fossil fuel policy was legitimized by an idea of "global cost-effectiveness": The core of the argument, as used in the Norwegian parliament, is that Norwegian oil and gas is cleaner (emission per unit produced) than fuel produced in other countries and if Norway does not produce to meet global demand, it will be replaced by fuels produced by other countries with higher emissions from production. Reducing Norwegian production and export would not also be very costly because it would reduce state income drastically. It was therefore argued to be the least cost-effective mitigation measure available, and it could potentially cause global emissions to increase. The quote below is from a parliamentary debate in 1997 and illustrates how this approach also was used in order to justify why Norway should increase production and export of oil and gas to reduce GHG emissions globally.

> *[…] increased Norwegian natural gas export will have a positive environmental effect when it replaces coal, and also increased Norwegian oil production could contribute to reducing global emissions if Norwegian production displaces less environmentally friendly production with higher greenhouse gas emissions per produced unit (Minister of Petroleum and energy Ranveig Frøiland (Labour Party), February 20, 1997; Verbatim report from the Storting, 1996–97, p. 2475 [authors' translation]; The Storting 1996–1997. Verbatim reports, n.d.).*

This policy position had broad support in the Norwegian Parliament for more than 15 years, from 1995 until approximately 2011. Andersen (2017) shows that this way of justifying further expansion of oil and gas activities at first gained broad support and legitimacy mainly because of two processes. First, although this way of arguing was used already in 1990, it first gained legitimacy after the establishment of new macroeconomic models that made it possible to calculate the cost-effectiveness for different mitigating measures in Norway. Although these calculations can be contested on a scientific basis (Fæhn et al., 2017), they were rarely contested in the Norwegian policy debates. In the debate these models were given epistemic authority, they were considered "neutral" and scientific, supporting a rational choice between various mitigation measures. Second, in the international arena Norway lobbied and worked to establish a global emission market where emission permits and projects that reduced emissions in other countries could be subtracted from national emissions (see Anker, 2018 for a summary of this development). Further, an international emission market was considered important to achieve the low-cost reductions. When it was implemented, it was taken as an international acceptance of the national policy by the majority in the Norwegian parliament. The principles of international emission trading were eventually enshrined in the 1997 Kyoto Protocol (Meckling, 2011). Both the development of "scientific" economic models and international climate negotiations increased the legitimacy for "global cost-effectiveness" approach in the Norwegian parliament. This effectively silenced most of the climate-based critique of Norwegian oil and gas production and exporting policies until 2011.

Over the past decade, however, there has been an intensification of the political debate on oil and gas policies. Rather than an abrupt change, this has been a gradual process that can be linked to several partly interwoven processes.

The idea of global cost-effectiveness has gradually been weakened because a comprehensive, global system for emission permits, for example, in the form of a global cap-and-trade system, did not emerge. Rather, regional and national systems evolved, and particular relevant for Norway is the establishment of a European Union Emission Trading Scheme (EU-ETS). In this system there is a cap on emissions and a market has been created to reach the goal of emission reduction at the lowest possible cost. This system has shifted the focus from global to European cost-effective reductions. EU-ETS has been important for the continued justification of the oil and gas export policies in terms of cost-effectiveness. However, increasing global emissions, various IPCC reports on the probability of severe societal consequences, and new developments in international negotiations have led to an increased urgency in the climate policy debate, and a subsequent emphasis on the need for reducing GHG emissions nationally. Simultaneously, Norwegian emissions have continued to rise, to a large degree because of the expanding oil and gas sector. By contrast, many other European countries have successfully been able to reduce their emissions.

The 2013 Parliamentary election marked a breakthrough for the Green party, and the party gained additional support in the 2017 election. Although small, the party has used parliamentary debates to argue for a managed draw-down of Norwegian oil and gas production in the coming decade, based on the point that Norway is exporting emissions (Bang & Lahn, 2019). The breakthrough of the Green party has also renewed similar critique from other political parties that used to argue for a less expansionist oil and gas policy.

Importantly, the Paris agreement in 2015 and development in climate science has fueled several new developments in public debates on Norwegian fossil fuel production and export. In particular, the claim that Norwegian production is substantially less emission intensive than in other countries has been contested in both public and parliamentary debates. The argument is still used but is now typically modified to "among the cleanest," a claim supported by a 2018 study published in Science (Masnadi et al., 2018).

In sum, the silence on the issue until 2011 has been replaced by an explicit debate on a "managed decline" of Norwegian production and export. The Paris agreement led the government to appoint a commission on green competitiveness and develop a white paper on how the Norwegian economy could succeed in a world with a stricter climate policy (MOF, 2017, 2018). The reports fueled a debate on the need to change the way the Norwegian government regulates and invests in the oil and gas sector. In this debate the economic risks of current oil and gas policy are used to justify a managed decline of the sector. Lahn and Bang (2019) identify a broad set of actors problematizing oil and gas tax policy and economic risks. This way of arguing supports a gradual change of current policies, "seeking to redefine rather than fundamentally oppose existing state interests in petroleum resource management, hence moving the understanding of oil from its current association with welfare towards one of (primarily economic) risk" (ibid.:1007). Further, rather than replacing the idea of global cost-effectiveness, this approach builds on a rationale in which the economic risks of current policies are problematic and potentially not in the interest of the people. The new language of risk is inspired by the international discussion of "stranded assets" in fossil fuels, where energy transition policies may cause investment in fossil fuel extraction to lose value before the end of its anticipated useful life (Caldecott, 2017). This represents a new kind of economic argument for a more restrictive fossil fuel policy that potentially could appeal to segments of the public that are less concerned about the GHG emissions from Norwegian oil and gas exports. Although debated, there is no official policy on a managed decline. In the latest white paper from spring 2021, the government outlined how it intend to meet its climate targets by 2030 (KMD, 2021). The main policy instrument is to gradually quadruple taxes on GHG emissions, from 590 NOK in 2021 to about 2000 NOK (ca. 240 USD) per tonne CO_2 equivalent by 2030. This will make it more expensive to produce oil and gas, and the Norwegian (and partly state owned) oil and gas company Equinor has launched plans to electrify many of the platforms to reduce emissions from production (Equinor, 2021).

Greater attention has also been directed toward the total amount of carbon that can be emitted if global warming is to be kept below 2°C. The carbon budget and the idea of a finite boundary of total emissions weaken "global cost-effectiveness" as a relevant way of assessing fossil fuel policies. This is because it is possible to have a very cost-effective climate policy that does not succeed in meeting the carbon budget. Following this approach, it has also been argued in Parliament and public debates that it is more or less irrelevant that Norwegian oil and gas are produced with very low emissions, because emissions from production only accounts for about 3% of the total emissions from burning oil and gas. Thus, the carbon budget approach can be used to justify the idea that a managed decline of oil and gas extraction is needed to stay within the carbon budget and reach the targets of the Paris agreement.

The economic risk and carbon budget perspectives have also often been combined in many policy debates. In these debates, a moral argument referring to future generations and division of wealth is used. The quote below is from a Parliamentary proposal in 2013 where the Green Party proposed to stop further oil and gas exploration in the Barents Sea in the Arctic. After referring to the carbon budget and the economic risks involved, the representative concluded:

> *[...] there are strong ethical arguments that Norway, as a country that has made a lot of money from fossil energy for many decades, should have a smaller share than poorer countries that have not started commercial operation of their oil and gas resources. It will be difficult for us to argue that Norway's share of the carbon budget should be larger than other countries' shares. This means that even in a very optimistic scenario where global coal production falls dramatically, significant parts of already known oil and gas reserves on the Norwegian shelf will have to remain under ground (Proposal from Rasmus Hansson (Green Party); The Storting, Dokument 8:39 S (2013–14), p. 2 [authors' translation]; The Storting 2013–2014 Dokument 8: 39, n.d.).*

This is just one example from a series of debates where economic risks, the carbon budget, and various moral risk arguments are combined to criticize Norwegian oil and gas production and export. The GHG emissions stemming from burning the energy that Norway exports are important in this critique. However, the political majority still clearly supports extending the oil and gas sector as long as possible. The observed changes are mainly discursive and have to a small extent resulted in changes in actual policy. However, Lahn and Bang (2019) argue that the policy positions that are critical toward further expansion of the sector are backed by a broad policy coalition that are problematizing oil and gas tax policy and economic risk. Their main point is that this coalition is gaining momentum and seems to incrementally change current policies by redefining the oil and gas sector: From a provider of welfare and wealth to a sector that poses (predominantly economic) risks. Even though there are few signs of an explicit plan to initiate a managed decline of the sector, there are trends in the policy debates that indicate that support for this is increasing.

Results from public opinion studies

While the political and public debate on Norwegian oil and gas exports has demonstrated both continuity and change over the decade since 2011, it is worth asking how public opinion has developed in this period. To further inform the findings from the policy domain, this section presents results of public opinion on issues related to the future of oil and gas export and connects it to climate change.

Public opinion on drilling in new areas

Fig. 7.2 displays trends in public opinion on drilling in the Lofoten, Vesterålen, and Senja archipelago from 2014 until 2020. Here we display the share of respondents placing themselves on the protection side of the scale.[2] Overall, the data show a majority against drilling in all waves. The share of respondents expressing opposition to drilling hovered just above half in the waves 2014–16. In the 2017–20 waves, the majority increased to about 60% expressing opposition to drilling.

[2] As mentioned in the methods section, the original scale goes from 1—strongly agree to 7—strongly disagree, with 4 representing a neutral neither/nor option.

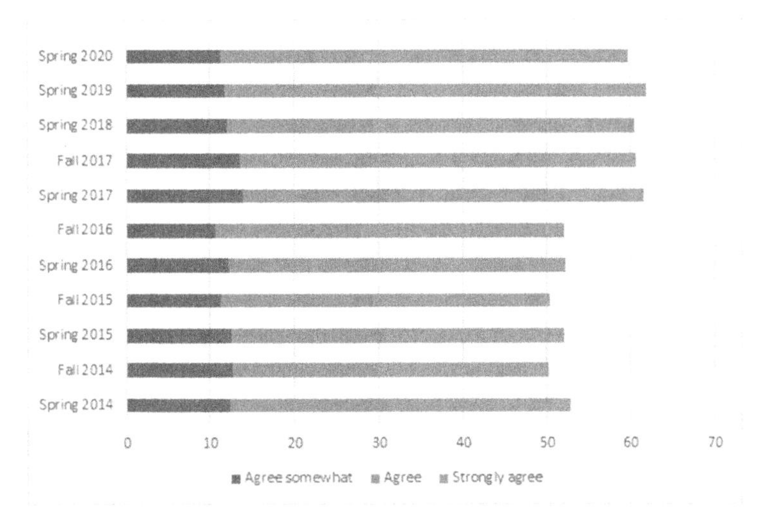

Figure 7.2 Trends in public opinion on oil and gas drilling in the Lofoten/Vesterålen/Senja archipelago, 2014–20. Percentage of respondents who agree that the area should not be opened for drilling. Disagreement and neutral responses not shown for clarity. Unweighted data.

It is noteworthy that most of the increase in support for protection over extraction is made up of an increase in the share of respondents agreeing "strongly" that the area should not be opened for drilling. Specifically, while the share of respondents choosing this option was seen in the low 20s in the first few years from 2014 on, the number hovered around the 30% mark starting from 2017. The break occurring in early 2017 divides the sample into two relatively stable periods, within each of which little change is seen. As we will explain in more detail in the Discussion section, we consider this change to be a result of increased public debate. Earlier work has shown this 2016–17 break to be statistically significant (Gregersen & Tvinnereim, 2020).

Future size of the oil and gas industry

Fig. 7.3 displays changes in how Norwegians see the future of the petroleum industry in a 20–30 years' perspective. Here our analyses is based on two data points as the question has only been asked twice. In 2016 and 2019, the modal response was that the size of the industry should be the same as the current level. This indicates stability. However, the share of respondents stating that the industry should be "considerably smaller than today" grew markedly, from 16.4% in 2016 to 23.8% in 2019, while the modal "maintained at its current level" option fell from 39.0% to 32.2%.

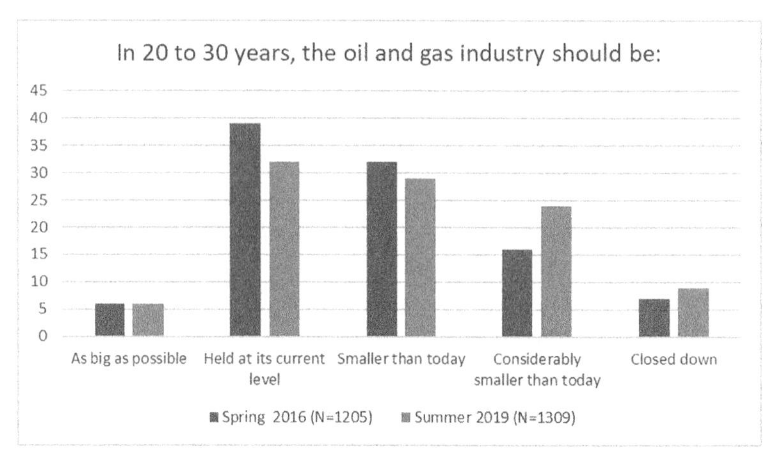

Figure 7.3 Trends in public opinion on the long-term size of the oil and gas industry. Data from 2016 to 2019, Norwegian Citizen Panel waves 6 and 15. N = 1205 (2016) and 1309 (2019). Unweighted data.

Is this change statistically significant? Assuming equal distances between the five response options, we can calculate the means in each year. Assigning the value of one to the most expansive option ("as big as possible") and five to the most restrictive ("closed down"), this yields an average score of 2.80 in 2016 and 2.97 in 2019. The observation counts are 1205 and 1,309, respectively, in the 2 years. The difference, while not particularly large, is statistically significant using a t-test (difference = 0.17, standard error of the difference = 0.042, t-statistic = −4.1). One worry might that the two samples are different, and that the difference seen between the 2 years derives from changes in the sample. Thus, we check the distribution for the 116 respondents in the panel who have answered the question in both years. The change in this group is similar to the previous test, at 0.17, with a standard error of 0.08 and t-statistic of 2.1. Furthermore, a chi-squared test of the response counts across the 2 years yields a score of 30, with four degrees of freedom, which results in P-value below 1%. The change in a more restrictive direction is thus statistically significant based on both balanced and unbalanced samples.

Worry about climate change

Fig. 7.4 displays trends in people's worry about climate change in Norway across 11 data points from 2013 to 2020. Overall, people do worry about climate change, as close to half select the options "worried" or "very

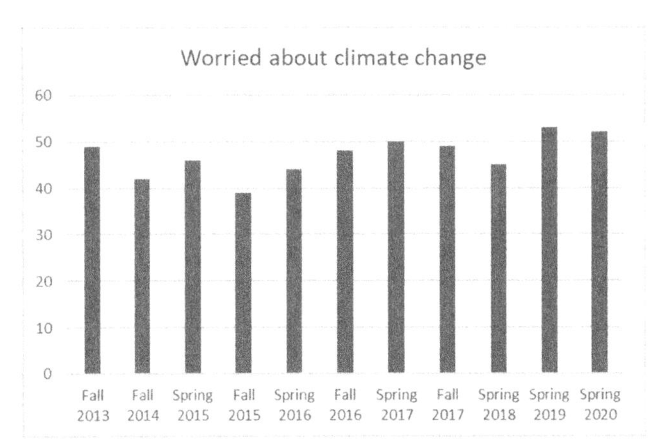

Figure 7.4 Worry about climate change. Shares of respondents selecting the options "worried" or "very worried" on a five-point scale. Data from 2013 to 2020, Norwegian Citizen Panel.

worried" across the years. However, the picture has not changed much from 2013 to 2020, signifying stability of opinions on this matter.

Discussion

At the beginning of this chapter, we set out to describe Norwegian public debate on the country's oil and gas exports and to analyze trends in public opinion related to such exports in the context of climate change. The material presented here allows us to make a few points.

First, the Norwegian debate on oil and gas exports is marked by a substantial degree of abstraction. The fact that oil and gas extraction takes place out of sight of shore removes the immediacy of extraction seen in so many other cases described in this volume. This makes the threat of oil spills and other environmental threats relatively remote—what is out of sight remains out of mind. Granted, many Norwegians work in the oil industry, and rig workers certainly experience the physical reality of fossil oil and gas. Nevertheless, most of the public has very little direct experience with petroleum installations. This lack of tangibility of oil and gas extraction, to most people, may in part explain the divergence between aggregate support for maintenance of the industry and opposition to new drilling around the Lofoten, Vesterålen, and Senja archipelago. The specific idea of drilling relatively near shore and next to important fishing grounds appears to reduce support relative to the generic idea of maintaining an established industry far offshore.

Second, attention to climate change is evident both in political debate and public opinion, followed by limited but still significant reduction in public support for continued production and export of oil and gas, particularly after 2016. The oil price crash in 2014, which hit the Norwegian economy hard in 2015 and 2016, notably with high unemployment in affected regions, reminded Norwegians of the vulnerability of being a small country that relies heavily on one industry in a downturn. Such a reaction is not a given, however, as an equally plausible response to the oil price decline could have been a redoubling of drilling efforts, notably if the industry did not display a concern about climate change. However, Tvinnereim and Ivarsflaten (2016) show that people who work in the Norwegian oil and gas industry in general support climate policies in the energy sector as much as the overall population, to the extent that such policies do not threaten their own employment opportunities. Thus, petroleum workers support carbon capture and storage, offshore wind power, and geothermal energy expansion as much as the general population, but less so emission and production restrictions. The adverse effect of the oil price decline of 2014 and the subsequent employment shock in the sector may thus have contributed to the slight decline in expectations of future importance and in support for new areas for extraction. This may have taken place in the context of an increased sense of the need for an energy transition—from fossil to renewable energy—although the concept of energy transition itself is not often used by the Norwegian public (Tvinnereim et al., 2020).

IPCC reports in 2013 and 2014 and the Paris Agreement of 2015 also spurred increased debate in Norway from 2016 and onwards. An emerging topic in the wake of the Paris agreement was whether the country needed a new oil policy that would limit exposure to the economic risk of relying on fossil fuel exports. The idea of global cost-effectiveness has also over time faded as a sufficient justification for allowing the petroleum industry to expand while claiming to address climate change. Notably, in 2016, a series of media stories and debates concluded that Norwegian oil is not the "world's cleanest" and that Norwegian oil and gas exports in practice entail the export of emissions to strengthen an already very strong economy even further. Thus, while there are few substantial changes in oil and gas policy over the 2013–18 period, a real political debate about central aspects has been raised for the first time in decades (Bang & Lahn, 2019). The brunt of these events take place in the 2014–16 period and may thus contribute to explaining the small but significant changes in public opinion. This may particularly be the case for the break seen between 2016 and 2017 in the question of oil drilling near the Lofoten, Vesterålen, and Senja archipelago.

A particularly interesting feature of the debate is the emerging realization of a paradox in Norwegian policy: The country is actively engaging in a policy to reduce demand for fossil energy, even though the success of such a policy would reduce the value of the country's economically most important export sector. The extensive domestic program to phase in electric vehicles is only one example of such a policy. This fact demonstrates the difficulty facing Norway in trying to balance different goals—that of general economic welfare and that of global climate protection. This tension is likely to grow over the coming decade.

References

Al-Kasim, F. (2006). *Managing petroleum resources: The 'Norwegian model' in a broad perspective*. Oxford Institute for Energy Studies.

Andersen, G. (2017). *Parlamentets natur. Utviklingen av en legitim miljø- og petroleumspolitikk (1945–2013)*. Scandinavian University Press. https://doi.org/10.18261/9788215028132-2017

Anker, P. (2018). A pioneer country? A history of Norwegian climate politics. *Climatic Change, 151*(1), 29–41. https://doi.org/10.1007/s10584-016-1653-x

Asdal, K. (2014). From climate issue to oil issue: Offices of public administration, versions of economics, and the ordinary technologies of politics. *Environment and Planning A, 46*(9), 2110–2124. https://doi.org/10.1068/a140048p

Bang, G., & Lahn, B. (2019). From oil as welfare to oil as risk? Norwegian petroleum resource governance and climate policy. *Climate Policy*, 1–13. https://doi.org/10.1080/14693062.2019.1692774

Caldecott, B. (2017). Introduction to special issue: Stranded assets and the environment. *Journal of Sustainable Finance & Investment,* 7(1), 1–13. https://doi.org/10.1080/20430795.2016.1266748

Dale, B., & Andersen, G. (2018). In H. Haarstad, & G. Rusten (Eds.), *Til Dovre faller?* (pp. 27–44). Scandinavian University Press.

Energy Facts Norway. *Electricity production* (2021). Norwegian Ministry of Petroleum and Energy https://energifaktanorge.no/en/norsk-energiforsyning/kraftproduksjon/.

Equinor. (2021). *Electrification of oil and gas operations*. Equinor. https://www.equinor.com/en/what-we-do/electrification.html.

Fæhn, T., Hagem, C., Lindholt, L., Mæland, S., & Rosendahl, K. E. (2017). Climate policies in a fossil fuel producing country – demand versus supply side policies. *The Energy Journal, 38*(1), 77–102. https://doi.org/10.5547/01956574.38.1.tfae

Gregersen, T., & Tvinnereim, E. (2020). *Olje- og gassutvinning i Lofoten og Vesterålen* (p. 2020). Energi Og Klima. https://energiogklima.no/nyhet/olje-gass-lofoten-vesteraalen/ (Original work published 2020).

KMD. (2021). *Klimaplan for 2021–2030 Meld. St. 13 (2020–2021)* (Whitepaper on climate action plan 2021–2030). Ministry of Climate and Environment https://www.regjeringen.no/no/dokumenter/meld.-st.-13-20202021/id2827405/.

Masnadi, M. S., El-Houjeiri, H. M., Schunack, D., Li, Y., Englander, J. G., Badahdah, A., Monfort, J. C., Anderson, J. E., Wallington, T. J., Bergerson, J. A., Gordon, D., Koomey, J., Przesmitzki, S., Azevedo, I. L., Bi, X. T., Duffy, J. E., Heath, G. A., Keoleian, G. A., McGlade, C., ... Brandt, A. R. (2018). Global carbon intensity of crude oil production. *Science, 361*(6405), 851–853. https://doi.org/10.1126/science.aar6859

Meckling, J. (2011). *Carbon coalitions: Business, climate politics, and the rise of emissions trading.* MIT Press.

MOF. (2017). *Perspektivmeldingen 2017. Meld. St. 29 (2016-2017)* (White paper on long-term economic perspectives). Ministry of Finance https://www.regjeringen.no/no/dokumenter/meld.-st.-29-20162017/id2546674/.

MOF. (2018). *Klimarisiko og norsk økonomi. Rapport fra klimarisikoutvalget* (Climate risk and the Norwegian economy. Report from the Climate Risk Commission). (NOU 2018:17). Ministry of Finance https://www.regjeringen.no/en/dokumenter/nou-2018-17/id2622043/.

NBIM. (2021). *Government pension fund global.* Norges Bank Investment Management. https://www.nbim.no/the-fund/.

Nilsen, Y. (2001). En felles plattform? Norsk oljeindustri og klimadebatten i Norge fram til 1998. *Acta Humaniora*, 267. University of Oslo.

Norwegian Petroleum. (2021). *Exports of oil and gas* (Vol. 2020). Ministry of Petroleum and Energy and the Norwegian Petroleum Directorate. https://www.norskpetroleum.no/en/production-and-exports/exports-of-oil-and-gas/.

Ramírez-Cendrero, J. M., & Wirth, E. (2016). Is the Norwegian model exportable to combat Dutch disease? *Resources Policy, 48*, 85–96.

Ryggvik, H. (2018). Norwegian oil workers: From rebels to parters in the tripartite system. In T. Atabaki, E. Bini, & K. Ehsani (Eds.), *Working for oil: Comparative social histories of labor in the global oil industry* (pp. 99–130). Springer International Publishing. https://doi.org/10.1007/978-3-319-56445-6_5

The Storting 1996–1997. Verbatim reports. (n.d.). Stortinget.

The Storting 2013-2014 Dokument 8: 39. (n.d.). Stortinget.

Tvinnereim, E., & Ivarsflaten, E. (2016). Fossil fuels, employment, and support for climate policies. *Energy Policy, 96*, 364–371. https://doi.org/10.1016/j.enpol.2016.05.052

Tvinnereim, E., Lægreid, O. M., & Fløttum, K. (2020). Who cares about Norway's energy transition? A survey experiment about citizen associations and petroleum. *Energy Research & Social Science, 62*, 101357. https://doi.org/10.1016/j.erss.2019.101357

von Brasch, T., Hungnes, H., & Strøm, B. (2019). *Ringvirkninger av petroleumsnæringen i norsk økonomi. Basert på endelige nasjonalregnskapstall for 2016 og 2017.* In Reports. Statistics Norway https://www.ssb.no/nasjonalregnskap-og-konjunkturer/artikler-og-publikasjoner/_attachment/405655?_ts=16ecb1da138.

CHAPTER 8

A "thin green line" of resistance? Assessing public views on oil, natural gas, and coal export in the Pacific Northwest region of the United States and Canada

Shawn Hazboun[1] and Hilary Boudet[2]
[1]Graduate Program on the Environment, The Evergreen State College, Olympia, WA, United States;
[2]Sociology, School of Public Policy, Oregon State University, Corvallis, OR, United States

Introduction

Growing awareness of the threat of global climate change has prompted nations around the world to seek various modes of decarbonization. In the United States and Canada, the widespread retirement of coal-fired power plants over the last decade (Burney, 2020; Storrow, 2019) has accounted for the majority of reductions in carbon emissions (EIA, 2019c; The Center Square, 2019), though these have largely been replaced with natural gas–fired power plants, rather than renewable energy facilities (Burney, 2020; Government of Canada, 2019b; Government of Canada, 2019c). The remaining coal plants in the Pacific Northwest region of the United States will be phased out in the near future, and across Canada a full phase-out of coal is planned for 2030 (Government of Canada, 2018). The declining demand for coal has meant that energy companies are increasingly seeking to establish trade routes with overseas markets, and have tirelessly proposed to build new export terminals in ports towns along the West Coast.

Simultaneously, the "shale revolution" of the late 2000s was precipitated by technological advances in extracting oil and natural gas from previously unreachable geologic formations (Wang et al., 2014), and by development of the Canadian oil sands. This "revolution" brought cheap and abundant natural gas to market in unprecedented amounts, and also resulted in levels of oil production that have not been seen in decades (EIA, 2019b; The Center Square, 2019).

Public Responses to Fossil Fuel Export
ISBN 978-0-12-824046-5
https://doi.org/10.1016/B978-0-12-824046-5.00003-5

The coastal towns of Pacific Northwest region of the United States and Canada are ideal export locations due to their existing deepwater ports and proximity to Western energy fields. Yet, this region is also known for its complex environmental politics, which in some locales are highly progressive and showcase strict environmental policy standards and an environmentally minded citizenry—but have also hosted environmental culture wars, such as the timber wars of the 1980 and 1990s over the protection of the endangered spotted owl (Dietrich, 1992). As such, proposals for new or expanded fossil fuel export facilities in this region have met fierce social opposition (Boudet, Baustad et al., 2018; Tran et al., 2019), especially when compared to energy infrastructure-dense areas like the Gulf Coast states of Texas and Louisiana, where state policies, existing infrastructure, and public opinion often facilitate new energy projects. The Pacific Northwest has been dubbed the "Achilles heel" of the energy industry (Grossman, 2019), standing between the abundant energy fields of the interior Western states of the United States and Canada and lucrative overseas markets, especially in Asia. A growing regional social movement seeks to block such infrastructure, and frequent protests, antifossil fuels campaigns, and the enactment of ordinances blocking movements of fossil fuels through municipal and county borders has helped the Pacific Northwest to earn another label, the "thin green line" (Washington State University - Vancouver, 2020). The "thin green line" concept is positioned as a call to action by regional environmental groups, urging activists and the broader public to stand against the ongoing efforts of energy companies to transform the region into a major hub for energy export (Sightline Institute, n.d.). Further contextualizing the "thin green line" is the growing global movement which seeks to constrain the global supply of all fossil fuels and proposes that remaining fossil fuel reserves should be "left in the ground" (Erickson et al., 2018; Piggot, 2018).

While regional and local environmental groups in the Pacific Northwest are actively engaged on the issues of obstructing fossil fuel production and export, the perception of the general public has not been widely studied. Yet, the expansion of energy exports in the Pacific Northwest—and how communities are impacted—will depend on the policy responses of state and local governments, the leverage of environmental social movements, and levels of public opposition or support in host communities and more broadly related to policymaking. This chapter assesses the state of public opinion across the region as a whole—the US states of Washington and Oregon and the Canadian province of British Columbia. We draw on data from a 2019 quota survey, which includes measures to assess public support/opposition to energy export, risk perceptions, familiarity with the issue, and individual characteristics.

We first discuss the state of fossil fuel export in the Pacific Northwest region, then review the literature covering public opinion on fossil fuel production and consider ways this may be similar or different to public opinion about fossil fuel export. We then describe our data and analytical strategy before presenting results and, finally, discussing the implications of our research for the region and the broader global fossil fuel trade.

Fossil fuel production, export, and policy in the Pacific Northwest

Despite increasing public pressure to reduce greenhouse gas emissions by moving away from fossil fuels–based energy, the United States and Canada have both been slow to make this transition, especially compared with many other developed countries such as those in West Europe, especially those that do not have their own fossil fuel resources (Karapin, 2020). Despite the significant phase-out of coal-fired electricity plants, there remains a continued heavy reliance on other fossil fuel for the electricity and transportation sectors, in part due to the availability of cheap natural gas and abundance of oil after the "shale revolution" of the late 2000s. Indeed, it is ironic that as the climate movement has gathered momentum and as renewable energy is proliferating, both natural gas and oil production are at an all-time high in both countries (EIA, 2018a,b, 2019a).

The Pacific Northwest region—conceptualized here as the US states of Washington and Oregon, and the Canadian province of British Columbia—are known as environmentally progressive places. This region is at the leading edge of climate and energy policymaking, including emphasis to reduce the use of fossil fuels for electricity production (though this is certainly made easier by the region's significant hydroelectric resources). In 2016, Oregon passed legislation to double the state's Renewable Portfolio Standard to 50% and has committed to retire all coal-fired electricity by 2030 (Gray & Bernell, 2020). Washington's governor recently created policy that aggressively reduces carbon emissions and sets a goal of 100% "clean" electricity by 2045 (Inslee, 2019). In British Columbia, policy requires 93% of all electricity to come from "clean or renewable sources" (BC Ministry of Energy and Mines, n.d.), and the province has even surpassed that level. Furthermore, individuals living in these states/province tend to be more likely to believe global warming is mostly caused by humans than in other parts of the United States and Canada, and also tend to be more likely to support climate policy (Marlon et al., 2019; Mildenberger & Howe, 2018).

Yet, significant extractive industries persist in this region. Though coal mining ceased in Washington in 2006, and historically there has been little to no oil or natural gas production, there are five major crude oil refineries in the state that together make Washington fifth in the nation for oil refining capacity, much of which is exported (EIA, 2019d). Oregon has no significant coal or oil production and just a few natural gas reserves, yet the state draws over 40% of its electricity from natural gas—fired power plants (compared with about 16% in Washington and less than 3% in British Columbia) (EIA, 2019c; The Center Square, 2019). Though British Columbia does not consume coal, coal mining is a major industry in the province, and British Columbia's coastal ports serve as major export terminals for coal produced there and in the Western United States (EIA, 2019b,e; Government of Canada, 2019a). Natural gas production is also a major industry in British Columbia, producing almost one-third of Canada's natural gas production in 2019 (Burney, 2020; Government of Canada, 2019c).

With the continued decline in coal consumption in the United States and Canada, as well as the abundance of oil and natural gas since the late 2000s, energy companies have become keenly interested in finding ways to expand export facilities to ship oil, gas, and coal to overseas markets. In the United States, export of oil became easier in early 2016 when President Obama lifted a 40-year ban on exporting crude oil. The United States is presently trending toward becoming a net exporter of oil and away from its current status as a net importer (EIA, 2018a,b, 2019a). The United States has been a net exporter of natural gas for 3 years in a row (EIA, 2020a,b). Currently, about 60% of US natural gas exports are conducted through pipeline and 40% as liquefied natural gas (LNG) on ships, which requires a special facility to compress the gas into a cooled liquid (EIA, 2020a,b). At present six LNG facilities are operating in four US states (Louisiana, Maryland, Georgia, and Texas), with several more either under construction or permitted to begin construction, including the highly controversial Jordan Cove facility on Oregon's coast (EIA, 2020b). Canada has just one operational LNG facility in New Brunswick (used only for import), though 13 LNG export terminals have been proposed in British Columbia alone and a major facility is currently under construction in Kitimat (Government of Canada, 2020).

In terms of coal, Western US coal states like Wyoming and Montana are currently bottle-necked for export options, and coal is currently shipped through the Seattle Customs District before it is sent north to export terminals in Vancouver, British Columbia. The lack of coal terminals along the west coast of the United States is not for a lack of effort on the part of coal companies—since 2010, seven major coal terminals have been proposed in Washington and Oregon. However, all proposals to date have failed due to a variety of permitting, legal, and social issues. Coal terminals exist in British Columbia, though several proposals to expand existing facilities or build new ones have also failed. Though it is not unusual in the realm of energy citing for a large proportion of energy infrastructure proposals to be met with opposition or to fail for other reasons (McAdam & Boudet, 2000), it is worth noting how much the fossil fuel industry has targeted this region of the United States and Canada in recent years and has been met with resistance, as compared with other regions such as the Gulf Coast.

Over the last decade, Oregon, Washington, and British Columbia have each hosted proposals for highly contentious fossil fuel export terminals and related pipelines and transportation routes (Sightline Institute, n.d.). Such proposals have usually been met with fierce opposition from environmental organizations, tribes and indigenous groups, policymakers, and the wider public. Several proposals have turned into high-profile battles, such as the Gateway Pacific Terminal, which would have exported Powder River Basin coal through a port just north of Bellingham, Washington. The controversy drew national attention and motivated several celebrities—including author and activist Bill McKibben—to get involved in the opposition effort (Rice, 2011), which eventually succeeded on the basis of tribal fishing rights (Allen et al., 2017).

Yet, host communities can be supportive of proposals for new fossil fuel export infrastructure, looking to the economic opportunities such facilities would bring. Additionally, while Native American tribes and Canadian First Nations often take opposed stances to fossil fuel projects, indigenous groups have also formed in support of these projects, such as the First Nations LNG Alliance in British Columbia (First Nations LNG Alliance, 2020).

Public opinion on fossil fuels: how does it relate to export?

North American energy social science, including public opinion research, has focused extensively (if not almost exclusively) on sites of fossil fuel extraction (mining, drilling), production (power plants, refining), and associated transportation facilities (pipelines, trains, trucking). For example, ample research exists on the socioeconomic and health impacts experienced by communities that host mining or power production facilities (Malin, 2015; Mayer et al., 2018), public perception about fossil fuel extraction technologies like hydraulic fracturing (Borick & Clarke, 2016; Boudet et al., 2014; Hazboun & Boudet, 2020), and social movement activity related to pipelines, oil and gas drilling on public lands, and so on (Boudet et al., 2017; McAdam & Boudet, 2000; Tran et al., 2019). Relatively less social science in the United States and Canada has examined the social dimensions of fossil fuel export, yet as mentioned this is of increasing importance in these countries. Little is known about how the public perceives the topic of fossil fuel export, and how highly related individuals' export and extraction/production attitudes are. There is reason to believe perceptions about export are distinct. For example, since export facilities are confined to coastal areas (compared with oil and gas wells or power plants), individuals might have a reduced sense of risk about their impacts. On the other hand, the most strongly associated predictors of individuals' attitudes about energy production are political ideology and political party affiliation (Boudet et al., 2014; Clarke et al., 2016; Gravelle & Lachapelle, 2015), and general environmental beliefs and global warming attitudes also tend to be strongly related to whether individuals support or oppose the production and consumption of fossil fuels (Ansolabehere & Konisky, 2009, 2014; Hazboun & Boudet, 2020; Mccright & Dunlap, 2011; McCright & Dunlap, 2011). So, individuals' attitudes may be more influenced by these factors than by risk perception or anticipation of economic benefit.

Two US studies have explicitly studied public opinion on fossil fuel export specifically. The first by Pierce et al. (2018) suggests that US citizens tend to be fairly supportive of increasing the export of natural gas to other countries. However, a regional study by Hazboun (2019) suggests that individuals in the Pacific Northwest region are more opposed than supportive of increased oil, coal, and natural gas export, though respondents expressed a high degree of uncertainty and unfamiliarity with the issue. Both studies found the correlates of supporting fossil fuel export to be similar to supporting fossil fuel extraction and consumption—including being male, politically conservative, and skeptical of anthropogenic global warming.

To investigate public views on fossil fuel export expansions in the Pacific Northwest region, we use data from a 2019 quota survey (n = 1500) of residents of Oregon, Washington, and British Columbia. We examine the relative level of support for expanded oil, natural gas, and coal export, and assess what factors help predict these views, including the relative influence of demographic characteristics, political ideology, environmental and climate change beliefs, local importance of fossil fuel industries, and familiarity with fossil fuel export. We also analyze the perceived benefits and risks of expanded fossil fuel export, as well as risk perceptions related to different modes of transportation including train, ship, and pipeline.

Methods: survey sampling and measurement

The authors contracted YouGov, a global public opinion vendor frequently used in public opinion research, to recruit the sample and administer the survey online. We chose the online survey format for its relative affordability (Campbell et al., 2018) but also because response rates to more traditional survey modes such as mail and telephone are increasingly subject to low response rates and resulting nonresponse bias (Groves, 2006; Stedman et al., 2019). High quality vendors like YouGov use quota sampling, sample weighting, and other strategies to ensure the nonprobability sample taken from their large respondent pools best approximates a probability sample, leading to more robust results.

We originally oversampled respondents from Oregon, Washington, and British Columbia, with equal amounts from rural and urban counties/districts, then reduced the sample to 1500 respondents using a matching and weighting procedure based on a sampling frame of gender, age, race, and education. This produced the final sample of 1500 respondents (500 from each place). YouGov then weighted the matched cases to the sampling frame using propensity scores. Last, YouGov calculated and included the sampling weights before presenting us with the final dataset.

Table 8.1 provides question wording, response options, and unweighted summary statistics for all variables used in the analysis. For the following multivariate regression analysis, our dependent variables are level of support for oil export, coal export, and natural gas export. We use a variety of independent variables to test the predictive association with the dependent variables across the three regression models, and these include sociodemographic characteristics (sex, age, race, education, residence in a metro or nonmetro county, state/providence of residence), political ideology, environment and climate change attitudes, local importance of energy industries, and familiarity with oil, coal, and natural gas export.

Table 8.1 Descriptive statistics for all variables (n = 1500).

Variable	Question wording and response options	Mean (std. dev.) or frequency (n)
Male	Are you male or female? (1) Male; (0) Female	42.2% (633) Male
Age	In what year were you born? (Subtracted from survey year, 2019). Range = 18 to 94	53.4 (16.2) years
White	What racial or ethnic group best describes you? (1) White; (2) Black; (3) Hispanic/Latino; (4) Asian; (5) Native American; (6) Mixed; (7) Other	84.5% (1267) White, non-Hispanic
Bachelors	What is the highest level of education you have completed? (1) 4-year college degree [bachelors]; (0) Less than 4-year college degree	40.1% (602) Bachelors
Metro	Respondents' county classified as metro or nonmetro	50% (750) Metro
Ideology	In general, how would you describe your own political viewpoint? (1) Very liberal; (2) Liberal; (3) Moderate; (4) Conservative; (5) Very conservative. *Variable collapsed to 3 categories for analysis*	43.9% Liberal, 28.4% Moderate, 27.6% Conservative
Env. v. Econ	With which one of these statements do you most agree? (1) Protection of the environment should be given priority, even at the risk of curbing economic growth; (0) Economic growth should be given priority even if the environment suffers to some extent	72.5% (1087) Environment
Anthro	Assuming global warming is happening, do you think it is: (1) Caused mostly by human activities; (0) Not caused mostly by human activities	51.7% (775) Human activities
Energy industries	Now, we'd like to ask you some questions about your local economy. Please indicate if [mining, refining, and utilities] is (1) not important at all; (2) a little important; (3) moderately important; (4) or very important to your area.	30.8% Not at all important; 38.3% A little important; 19.8% Moderately important; 11% Very important

Table 8.1 Descriptive statistics for all variables (n = 1500).—cont'd

Variable	Question wording and response options	Mean (std. dev.) or frequency (n)
Familiarity: Oil	How much have you ever heard or read about the United States/Canada exporting the following fossil fuels to other countries [OIL]. (1) Not at all; (2) A little; (3) A moderate amount; (4) A lot; (5) A great deal	18.7% Not at all; 26.7% A little; 28.1% A moderate amount; 17% A lot; 9.5% A great deal
Familiarity: Coal	How much have you ever heard or read about the United States/Canada exporting the following fossil fuels to other countries [COAL]. (1) Not at all; (2) A little; (3) A moderate amount; (4) A lot; (5) A great deal	32.8% Not at all; 28.5% A little; 23.0% A moderate amount; 10.1% A lot; 5.6% A great deal
Familiarity: Natural gas	How much have you ever heard or read about the United States/Canada exporting the following fossil fuels to other countries [NATURAL GAS]. (1) Not at all; (2) A little; (3) A moderate amount; (4) A lot; (5) A great deal	26.8% Not at all; 23.6% A little; 27.8% A moderate amount; 13.6% A lot; 8.2% A great deal
Support export: Oil	To what extent do you OPPOSE or SUPPORT the United States exporting the following fossil fuels to other countries? [OIL] (1) Strongly oppose; (2) Somewhat oppose; (3) Not sure; (4) Somewhat support; (5) Strongly support	34.8% Strongly oppose; 20.1% Somewhat oppose; 6.3% Not sure; 20.2% Somewhat support; 18.6% Strongly support
Support export: Coal	To what extent do you OPPOSE or SUPPORT the United States exporting the following fossil fuels to other countries? [COAL] (1) Strongly oppose; (2) Somewhat oppose; (3) Not sure; (4) Somewhat support; (5) Strongly support	39.3% Strongly oppose; 19.1% Somewhat oppose; 8.3% Not sure; 20.49% Somewhat support; 12.8% Strongly support
Support export: Natural gas	To what extent do you OPPOSE or SUPPORT the United States exporting the following fossil fuels to other countries? [NATURAL GAS (for example, as LNG)] (1) Strongly oppose; (2) Somewhat oppose; (3) Not sure; (4) Somewhat support; (5) Strongly support	24.8% Strongly oppose; 19.7% Somewhat oppose; 9.8% Not sure; 24.0% Somewhat support; 21.8% Strongly support

Unweighted variable means and proportions reported.

Though not shown in Table 8.1, we also measured respondents' perceptions of the risks and benefits of fossil fuel export by asking them to select from a provided list which was a potential benefit from the US exporting fossil fuels to other countries, and which was a potential risk from the US exporting fossil fuels to other countries. The provided list included eight items and was the same for the benefits and risks question: jobs in the region, global climate, energy security (reliable access to energy), regional environment (health of animals, plants, and their habitat), energy prices, regional economy (tax base, businesses, etc.), public health (air quality, pollution, etc.), private property (property values, eminent domain, etc.), and other (open-ended).

Last, we also measured respondents' relative risk perceptions related to the transportation of oil, coal, and natural gas by railroad, ship, and pipeline (not shown in Table 8.1). We asked respondents, "In general, do you think transporting the following fossil fuels by [railroad, ship, pipeline] is safe or unsafe [oil, coal, natural gas]?" and respondents answered on a five-point scale from Completely Safe to Extremely Unsafe, with an "Unsure" middle option.

Results

We first examine respondents' relative support for oil, coal, and gas export, using weighted summary statistics. Fig. 8.1 indicates that respondents were the most supportive of natural gas export, with about 48% indicating they were either moderately or extremely supportive. Conversely, respondents were the most opposed to coal export (57% opposed), followed by oil export (53% opposed). Additionally, respondents indicated the highest level of uncertainty (13%) about their support for natural gas export, compared with oil and coal export. Notably, while a majority of respondents were opposed to oil and coal export, only 39% were opposed to natural gas export.

Next, we assess perceptions of the risk and benefits of fossil fuel export. Fig. 8.2 indicates that the three most commonly perceived risk of fossil fuel export were impacts to the regional environment (60%), global climate (59%), and public health (61%). Conversely, the three most commonly perceived benefits were jobs in the region (67%), regional economy (55%), and energy prices (46%).

Fig. 8.3 displays respondents' risk perceptions related to different transportation modes for oil, coal, and natural gas. Overall, respondents

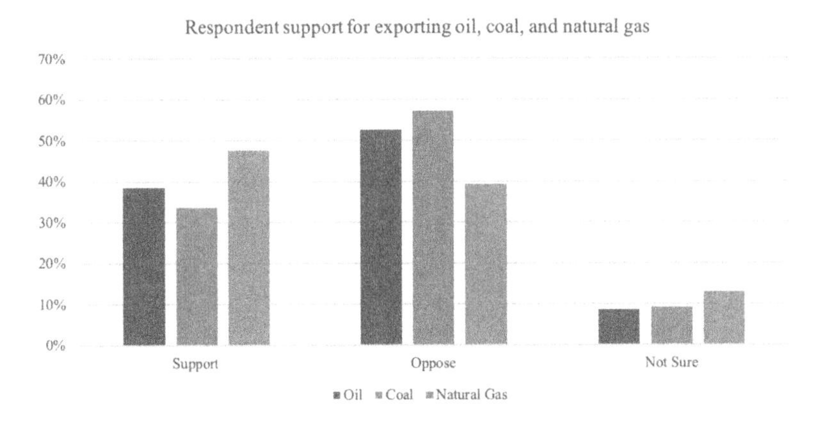

Figure 8.1 Respondents' support for oil, coal, and natural gas export, calculated using survey weights.

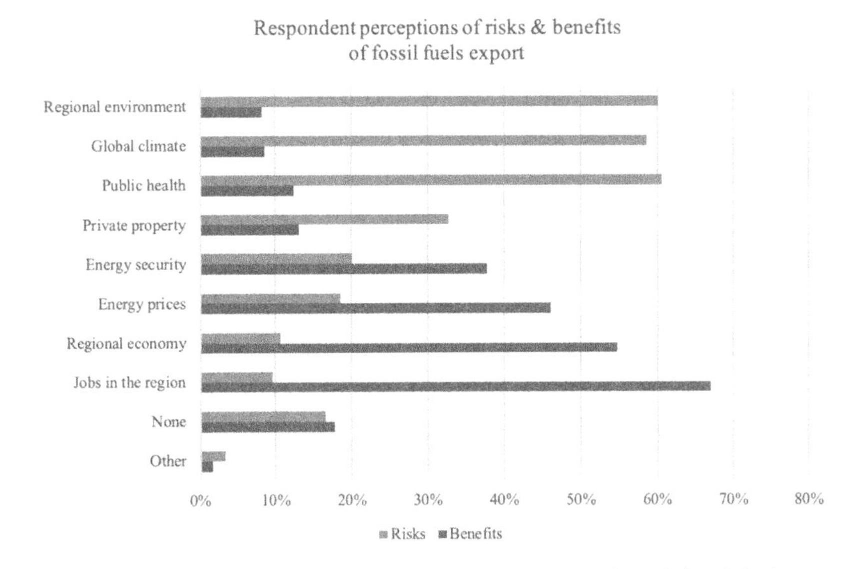

Figure 8.2 Respondents' agreement with risks and benefits of fossil fuel export, calculated using survey weights.

were the least concerned about transporting coal by railroad and ship (67% and 64% thought it was completely or mostly safe, respectively). Less than half of respondents thought transporting oil by railroad (48%) and ship (45%) was safe, as with transporting natural gas by railroad (45%) and ship (49%). As far as the safety of pipelines, about half of respondents thought transporting both oil and

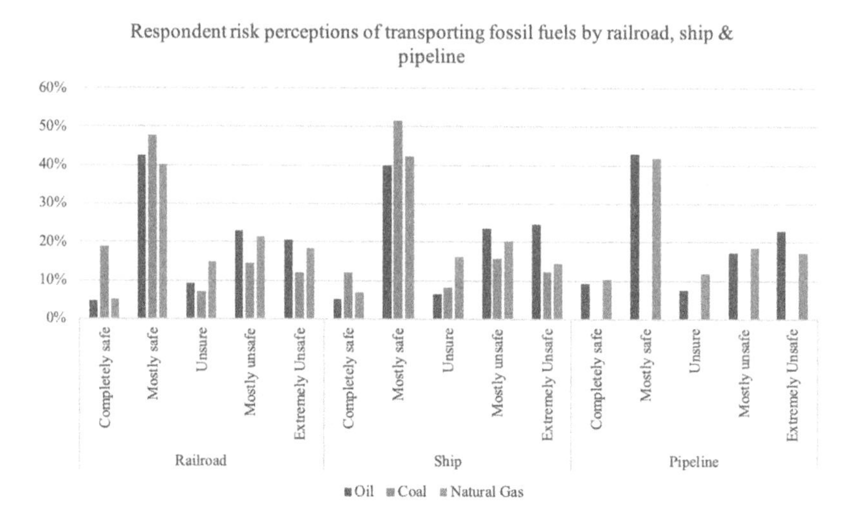

Figure 8.3 Summary statistics for respondent risk perceptions for transporting oil, coal, and natural gas by railroad, ship, and pipeline, calculated with survey weights.

natural gas by pipeline was completely or mostly safe, while the rest though it was unsafe or were unsure. The highest level of uncertainty about transporting any of the three fossil fuels was for natural gas.

We next estimate three regression models to analyze the factors that predict respondents' support for oil, coal, and natural gas export. Since the dependent variables were measured as categorical, we tested the proportional odds assumption for ordered logistic regression. However, this assumption was violated in all three models, so we collapsed the dependent variable to a "support" or "not support" measure and estimated binary logistic regression models. We used survey weights in each model (described in methods section). Table 8.2 provides the results of the regression models, and odds ratios are reported.

Across the three models, some similarities and differences stand out. First, being male is a statistically significant, positive predictor of supporting oil and natural gas (but not coal) export, all else being equal. However, sex is the only sociodemographic factor that has a significant association with the dependent variables. Place of residence is significant, with respondents from Washington having statistically lower odds of supporting oil, coal, and natural gas export than British Columbia respondents, and respondents from Oregon having lower odds of supporting oil and natural gas (but not coal). Political ideology was only statistically significant in the coal and

Table 8.2 Weighted binary logistic regression predicting support for oil, coal, and natural gas export.

	Oil	Coal	Natural gas
Male	1.686*	1.128	2.362***
Age	1.012	0.997	1.007
White	0.728	0.657	0.628
Bachelors	1.251	1.117	0.924
Metro	1.18	1.683	1.493
State (ref = BC)			
Washington	0.328***	0.541*	0.351***
Oregon	0.294***	1.005	0.365***
Ideology (ref = Liberal)			
Moderate	1.187	1.761*	1.973**
Conservative	1.753	3.426***	1.920*
Belief in AGW	0.198***	0.335***	0.335***
Env. v. Econ.	0.211***	0.311***	0.312***
Familiarity	1.266*	1.245*	1.370***
Energy industries	1.366*	1.276*	1.208
_cons	0.898	0.417	0.778
Model characteristics			
n	1377	1376	1376
Pseudo R-squared	0.337	0.261	0.265
AIC	1255	1336	1428

Odds ratios reported.*$P < .05$; **$P < .01$; ***$P < .001$.

natural gas models, in which the odds of support were 70% to three times as high for moderates and conservatives than for liberals.

As far as the attitudinal predictor variables, prioritizing the environment over the economy and believing in anthropogenic global warming were statistically significant, negative predictors of support for export of all three fossil fuels. However, familiarity with oil, coal, and natural gas export had a statistically significant but positive relationship with the dependent support variables. Last, local importance of energy industries (mining, refining, and utilities) was positively associated with support for both coal and natural gas export, but not oil export.

Turning to the metrics for model quality, both the pseudo R-squared value as well as the AIC values indicate that the first model—support for oil export—has the best fit, with the included predictor variables together accounting for about 34% of the variance in the dependent variable. The models for coal and natural gas export explain about 26% and 27% of the variance in the dependent variables, respectively.

Discussion, implications, and future research

The purpose of this research was to assess general public perceptions related to fossil fuel export, an increasingly significant topic in the Pacific Northwest region of the United States and Canada. Though Washington, Oregon, and British Columbia have all seen an increase in proposals from energy companies to expand export facilities or build new ones, this effort has been met with well-organized and widespread social opposition. Yet, public opinion in this region is not well studied. Using a 2019 online quota survey, we measured respondents' support for oil, coal, and natural gas export, perceptions of benefits and risks from fossil fuel export, risk perceptions related to different modes of fossil fuel transportation, level of familiarity with fossils export, and other relevant factors including political ideology and global warming beliefs.

We found that while a majority of respondents were opposed to oil and coal export, just 39% were opposed to natural gas export, while 48% were supportive and the rest uncertain. These results are in line with the findings of Hazboun (2019) and Pierce et al. (2018). We suggest this finding reflects not only uncertainty about the technological aspects of LNG but also a broader societal debate about the environmental merits of natural gas. Natural gas has not only been heralded as a "bridge fuel" to a renewable energy future but also as a less impactful fossil fuel for the climate since it emits less carbon than oil or coal and can offset carbon emissions from coal-fired power plants (Delborne et al., 2020; LNG Canada, 2020). When the shale revolution made natural gas extremely affordable and widely accessible, some major US environmental groups promoted it as the bridge away from coal. However, growing awareness about natural gas's environmental impacts, including its methane emissions, have sparked a debate about its appropriate role in the energy transition and caused the Sierra Club and other organizations to change their stance (Casselman, 2009; Walsh, 2012). Turning back to our findings, we expect respondents' relatively mixed attitudes about natural gas export are related to their more general mixed attitudes about natural gas itself. Future research could further explore the reasons why individuals might be more supportive of natural gas export compared with oil or coal export.

We also found that respondents were fairly concerned about the transportation processes for oil and natural gas, though less so for coal. This included transportation by train, ship, and pipeline, though as with the measures of support respondents indicated the most uncertainty relative to transporting natural gas. This could be a result of natural gas being perceived as more explosive than oil, which might offset more positive views of natural gas's potential as a bridge fuel. The technological aspects of LNG might also be less familiar to respondents, and though LNG technology is not new (it has been used to import natural gas for decades), it has been receiving media attention in recent years due to the flurry of new LNG export terminal projects in both the United States and Canada. Additionally, respondents' concern about oil transportation could possibly be explained by a number of well known, highly visible oil spills that have been covered extensively by the media over the years.

In our regression models, we found that sex, political ideology, climate change beliefs, environmental prioritization, familiarity, and local importance of energy industries to be significant predictors of support for export of two of the three fossil fuel. These findings are consistent with the literature on public attitudes about energy and environmental topics. We were somewhat surprised that being politically conservative was not a significant predictor of support for oil export; instead, other included independent variables were stronger predictors of support. We were interested to note variation by state/providence; across the three models, respondents from Oregon and Washington were less supportive of export than respondents from British Columbia. One possible explanation for this is that British Columbia, while progressive in terms of climate and energy policy, hosts major fossil fuel extraction operations and thus respondents from that province likely feel that expanding export capacity is crucial to the future stability of their employers and industry—several studies from the United States suggest that fossil fuel industry presence, affiliation, or proximity increase support for fossil fuels (Boudet et al., 2016; Boudet, Zanocco et al., 2018).

So far, very few proposed fossil fuel export terminals that have been proposed in the Pacific Northwest in the last 10 years have been permitted and built. Reasons why this is have ranged from significant social and community opposition, permitting failures after environmental review, and impacts to tribal fishing rights, among other factors. The growing regional and global movement against continued development of fossil fuel infrastructure is not likely to abate in coming years. Yet, the fossil fuel industry will continue to face mounting pressure to get their product to market. If industry succeeds in positioning the region as a fossil fuel export hub, activists contend this could be a sort of Pandora's box for the climate, unlocking further mobility and motivation for companies operating in Western energy fields to extract fossil fuels and ship them around the world. The future role of the Pacific Northwest in the global fossil fuel trade, and the strength of the "thin green line," will depend partly on the responses of public leaders, citizens, and organized groups as they continue to respond to proposals from industry.

References

Allen, M., Bird, S., Breslow, S., & Dolšak, N. (2017). Stronger together: Strategies to protect local sovereignty, ecosystems, and place-based communities from the global fossil fuel trade. *Marine Policy, 80*, 168–176. https://doi.org/10.1016/j.marpol.2016.10.019

Ansolabehere, S., & Konisky, D. M. (2009). Public attitudes toward construction of new power plants. *Public Opinion Quarterly, 73*(3), 566–577. https://doi.org/10.1093/poq/nfp041

Ansolabehere, & Konisky, D. M. (2014). *Cheap and clean: How Americans think about energy in the age of global warming.*

BC Ministry of Energy and Mines. (n.d.). *Renewable energy—province of British Columbia.* https://www2.gov.bc.ca/gov/content/industry/electricity-alternative-energy/renewable-energy.

Borick, C. P., & Clarke, C. (2016). American views on fracking (SSRN scholarly paper ID 2781503). *Social Science Research Network.* https://papers.ssrn.com/sol3/papers.cfm?abstract_id=2781503.

Boudet, H., Baustad, B., & Tran, T. (2018). Protest in the siting of liquified natural gas terminals in Oregon. In *Fractured communities: Risk, impacts, and protest against hydraulic fracking in U.S. Shale regions.*

Boudet, H., Bugden, D., Zanocco, C., & Maibach, E. (2016). The effect of industry activities on public support for 'fracking. *Environmental Politics, 25*(4), 593–612. https://doi.org/10.1080/09644016.2016.1153771

Boudet, H., Clarke, C., Bugden, D., Maibach, E., Roser-Renouf, C., & Leiserowitz, A. (2014). Fracking controversy and communication: Using national survey data to understand public perceptions of hydraulic fracturing. *Energy Policy, 65*, 57–67. https://doi.org/10.1016/j.enpol.2013.10.017

Boudet, H., Trang, T., & Gaustad, B. (2017). The long shadow of unconventional hydrocarbon development: Contentious politics in liquefied natural gas (LNG) facility siting in Oregon. In *Fractured communities: Risks, impact, and protest against hydraulic fracking in U.S. Shale regions. E.*

Boudet, H. S., Zanocco, C. M., Howe, P. D., & Clarke, C. E. (2018). The effect of geographic proximity to unconventional oil and gas development on public support for hydraulic fracturing. *Risk Analysis : An Official Publication of the Society for Risk Analysis, 38*(9), 1871–1890. https://doi.org/10.1111/risa.12989

Burney, J. A. (2020). The downstream air pollution impacts of the transition from coal to natural gas in the United States. *Nature Sustainability, 3*(2), 152–160. https://doi.org/10.1038/s41893-019-0453-5

Campbell, R. M., Venn, T. J., & Anderson, N. M. (2018). Cost and performance tradeoffs between mail and internet survey modes in a nonmarket valuation study. *Journal of Environmental Management, 210*, 316–327. https://doi.org/10.1016/j.jenvman.2018.01.034
Casselman, B. (2009). Sierra club's pro-gas dilemma. *Wall Street Journal*. https://www.wsj.com/articles/SB126135534799299475.
Clarke, C. E., Budgen, D., Hart, P. S., Stedman, R. C., Jacquet, J. B., Evensen, D. T. N., & Boudet, H. S. (2016). How geographic distance and political ideology interact to influence public perception of unconventional oil/natural gas development. *Energy Policy, 97*, 301–309. https://doi.org/10.1016/j.enpol.2016.07.032
Delborne, J. A., Hasala, D., Wigner, A., & Kinchy, A. (2020). Dueling metaphors, fueling futures: "Bridge fuel" visions of coal and natural gas in the United States. *Energy Research and Social Science, 61*. https://doi.org/10.1016/j.erss.2019.101350
Dietrich, W. (1992). *The final forest: The battle for the last great trees of the Pacific Northwest*.
EIA. (2018a). For one week in November, the U.S. was a net exporter of crude oil and petroleum products. *Today in Energy*. https://www.eia.gov/todayinenergy/detail.php?id=37772.
EIA. (2018b). U.S. liquefied natural gas export capacity to more than double by the end of 2019. In *Today in energy*.
EIA. (2019a). *Annual energy outlook 2019 with projections to 2050* (AEO 2019 ed., p. 83) https://www.eia.gov/state/analysis.php?sid=WA#115.
EIA. (2019b). *Oregon—state energy profile overview - US energy information administration*. https://www.eia.gov/state/?sid=OR#tabs-4.
EIA. (2019c). *U.S. Energy-related carbon dioxide emissions, 2018*. https://www.eia.gov/environment/emissions/carbon/.
EIA. (2019d). *U.S. Number and capacity of petroleum refineries*. https://www.eia.gov/dnav/pet/pet_pnp_cap1_dcu_nus_a.htm.
EIA. (2019e). *Washington—state energy profile analysis—U.S. energy information administration*. https://www.eia.gov/state/?sid=WA#tabs-4.
EIA. (2020a). *Liquefied natural gas*. U.S. Energy Information Administration (EIA).
EIA. (2020b). *Natural gas imports and exports*. U.S. Energy Information Administration. https://www.eia.gov/energyexplained/natural-gas/imports-and-exports.php.
Erickson, P., Lazarus, M., & Piggot, G. (2018). Limiting fossil fuel production as the next big step in climate policy. *Nature Climate Change, 8*(12), 1037–1043. https://doi.org/10.1038/s41558-018-0337-0
First Nations LNG Alliance. (2020). https://www.fnlngalliance.com/.
Government of Canada. (2018). *Canada's coal power phase-out reaches another milestone*. Gcnws. https://www.canada.ca/en/environment-climate-change/news/2018/12/canadas-coal-power-phase-out-reaches-another-milestone.html.
Government of Canada. (2019a). *National energy board, provincial and territorial energy profiles – British Columbia*. https://www.cer-rec.gc.ca/nrg/ntgrtd/mrkt/nrgsstmprfls/bc-eng.html.
Government of Canada. (2019b). *Natural resources Canada, electricity facts*. https://www.nrcan.gc.ca/science-data/data-analysis/energy-data-analysis/energy-facts/electricity-facts/20068#L1.
Government of Canada. (2020). *Natural resources Canada - Canadian LNG projects*. https://www.nrcan.gc.ca/our-natural-resources/energy-sources-distribution/clean-fossil-fuels/natural-gas/canadian-lng-projects/5683.
Government of Canada. (2019c). *NEB – market snapshot: Canada's power generation: Switching from coal to natural gas*. https://www.cer-rec.gc.ca/nrg/ntgrtd/mrkt/snpsht/2017/04-02cndpwrgnrtn-eng.html.
Gravelle, T. B., & Lachapelle, E. (2015). Politics, proximity and the pipeline: Mapping public attitudes toward Keystone XL. *Energy Policy, 83*, 99–108. https://doi.org/10.1016/j.enpol.2015.04.004

Gray, D., & Bernell, D. (2020). Tree-hugging utilities? The politics of phasing out coal and the unusual alliance that passed Oregon's clean energy transition law. *Energy Research and Social Science, 59*, 101288. https://doi.org/10.1016/j.erss.2019.101288

Grossman, Z. (2019). Native/non-native alliances challenging fossil fuel industry shipping at Pacific Northwest ports. In J. Clapperton, & L. Piper (Eds.), *Environmental activism on the ground: Small green and indigenous organizing* (pp. 47–72). University of Calgary Press.

Groves, R. M. (2006). Nonresponse rates and nonresponse bias in household surveys. *Public Opinion Quarterly, 70*(5), 646–675. https://doi.org/10.1093/poq/nfl033

Hazboun, S. O. (2019). A left coast 'thin green line'? Determinants of public attitudes toward fossil fuel export in the Northwestern United States. *Extractive Industries and Society, 6*(4), 1340–1349. https://doi.org/10.1016/j.exis.2019.10.009

Hazboun, S. O., & Boudet, H. S. (2020). Public preferences in a shifting energy future: Comparing public views of eight energy sources in North America's Pacific Northwest. *Energies, 13*(8). https://doi.org/10.3390/en13081940

Inslee, J. (2019). *Energy & environment*. Governor Jay Inslee.

Karapin, R. (2020). Federalism as a double-edged sword: The slow energy transition in the United States. *The Journal of Environment and Development, 29*(1), 26–50. https://doi.org/10.1177/1070496519886001

LNG Canada.(2020). https://www.lngcanada.ca/about-lng-canada/.

Malin, S. A. (2015). *The price of nuclear power: Uranium communities and environmental justice*.

Marlon, J., Howe, P., Mildenberger, M., Leiserowitz, A., & Wang, X. (2019). *Yale climate opinion maps*.

Mayer, A., Olson-Hazboun, S. K., & Malin, S. (2018). Fracking fortunes: Economic well-being and oil and gas development along the urban-rural continuum. *Rural Sociology, 83*(3), 532–567. https://doi.org/10.1111/ruso.12198

McAdam, D., & Boudet, H. (2000). *Putting social movements in their place: Explaining opposition to energy projects in the United States*.

McCright, A. M., & Dunlap, R. E. (2011). Cool dudes: The denial of climate change among conservative white males in the United States. *Global Environmental Change, 21*(4), 1163–1172. https://doi.org/10.1016/j.gloenvcha.2011.06.003

Mccright, A. M., & Dunlap, R. E. (2011). The politicization of climate change and polarization in the American public's views of global warming, 2001–2010. *The Sociological Quarterly, 52*(2), 155–194. https://doi.org/10.1111/j.1533-8525.2011.01198.x

Mildenberger, M., & Howe, P. (2018). *Canadian climate opinion maps*. https://climatecommunication.yale.edu/visualizations-data/ccom/.

Pierce, J. J., Boudet, H., Zanocco, C., & Hillyard, M. (2018). Analyzing the factors that influence U.S. public support for exporting natural gas. *Energy Policy, 120*, 666–674. https://doi.org/10.1016/j.enpol.2018.05.066

Piggot, G. (2018). The influence of social movements on policies that constrain fossil fuel supply. *Climate Policy, 18*(7), 942–954. https://doi.org/10.1080/14693062.2017.1394255

Rice. (2011). *Northwest coal port ignites controversy*. High Country News.

Sightline Institute. (n.d.). *The thin green line*. https://www.sightline.org/research/thin-green-line/.

Stedman, R. C., Connelly, N. A., Heberlein, T. A., Decker, D. J., & Allred, S. B. (2019). The end of the (research) world as we know it? Understanding and coping with declining response rates to mail surveys. *Society and Natural Resources, 32*(10), 1139–1154. https://doi.org/10.1080/08941920.2019.1587127

Storrow, B. (2019). And now the really big coal plants begin to close - scientific American. In *E&E news*. https://www.scientificamerican.com/article/and-now-the-really-big-coal-plants-begin-to-close/.

The Center Square, Bethany Blankley. (2019). *Report: Natural gas outpaced renewable energy in reducing greenhouse gases*. The Center Square. https://www.thecentersquare.com/national/report-natural-gas-outpaced-renewable-energy-in-reducing-greenhouse-gases/article_ba2b492a-1618-11ea-ab0f-3f53387abb18.html.

Tran, T., Taylor, C. L., Boudet, H. S., Baker, K., & Peterson, H. L. (2019). Using concepts from the study of social movements to understand community response to liquefied natural gas development in Clatsop County, Oregon. *Case Studies in the Environment, 3*(1). https://doi.org/10.1525/cse.2018.001800

Walsh, B. (2012). *How the Sierra Club took millions from the natural gas industry—and why it stopped*. Time. https://science.time.com/2012/02/02/exclusive-how-the-sierra-club-took-millions-from-the-natural-gas-industry-and-why-they-stopped/.

Wang, Q., Chen, X., Jha, A. N., & Rogers, H. (2014). Natural gas from shale formation – the evolution, evidences and challenges of shale gas revolution in United States. *Renewable and Sustainable Energy Reviews, 30*, 1–28. https://doi.org/10.1016/j.rser.2013.08.065

Washington State University - Vancouver. (2020). *The thin green line is people's history*. https://labs.wsu.edu/thethingreenlineispeople/.

PART IV

Community response to export projects

CHAPTER 9

Global discourses, national priorities, and community experiences of participation in the energy infrastructure projects in northern Russia

Julia Loginova
School of Earth and Environmental Sciences, The University of Queensland, St Lucia, QLD, Australia

Introduction

Russia is among the world's largest exporters of carbon dioxide emissions (CO_2) in fossil fuels (International Energy Agency, 2020). Contributing to these emissions are extensive energy infrastructure projects developed over the last 2 decades in northern Russia (Graybill, 2017). These projects include unconventional oil extraction, liquefied natural gas (LNG), and large-scale pipeline projects taking place in the Arctic, Siberia, and the Far East, primarily oriented at the export of fossil fuels. To understand the dynamics of fossil fuel export, it is important to explore the social dimensions of expanding fossil fuel production. There remains a paucity of evidence on responses in communities affected by resource extraction projects in Russia.

It is now well established that resource extraction projects should be accompanied by meaningful community participation, and a lack of participation increases the likelihood of community resistance (Conde & Le Billon, 2017). Here, I will argue that on-the-ground public responses are best understood when community participation practices are considered in a dynamic interaction with global discourses and national priorities for development. Discourses and practices are common dimensions of problematization of forms of knowledge and experiences as they help to shape the reality to which they refer (Jentoft, 2017). I aim to examine these multifaceted relations in the specific context of the northern Russian regions.

Public Responses to Fossil Fuel Export
ISBN 978-0-12-824046-5
https://doi.org/10.1016/B978-0-12-824046-5.00012-6

Drawing on the literature on resource industries (Bebbington & Bury, 2013) and using the power of exclusion as a conceptual framework (Hall et al., 2011), I focus on the exclusionary aspects of community participation: while communities supposedly participate in the decision-making process and in the benefits from fossil fuel projects, they are facing various challenges that limit meaningful community participation. The study identifies that exclusionary processes are driven by several strategies: discursive strategies, market mechanisms, legal and bureaucratic strategies, and strategies of uncertainty. These strategies constitute the exercise of the "hidden faces of power," which Gaventa defined as forces that shape actions in ways not apparent in formal processes (Gaventa, 1982).

This chapter is based on qualitative research methods. The data were collected through field research conducted in 2015 in rural and Indigenous communities in the Republic of Komi (Komi) (Shelyaur, Shelyabozh, Kolva) and the Republic of Sakha (Yakutia) (Khatystyr and Iengra). Data collection methods included semistructured interviews (n = 25), group discussions (n = 2), and document (consultation policies and guidelines, reports by NGOs and community organizations) and media analysis. Community insights illuminate the microdynamics at play, including the place-based experiences of community participation and community response. These local-level perspectives were linked to broader socioeconomic and geopolitical processes highlighted in semistructured interviews with company representatives (n = 14) and government officials (n = 22). In the next section, I briefly explain the shifting spatialities of fossil fuel projects in Russia. Then, I draw together existing research on community impacts and responses to energy projects from across northern Russia. Next, I draw on the empirical study and examine exclusionary strategies and relate them to the multiscalar processes at play in the development of energy projects in northern Russia.

Shifting spatialities of Russia's oil and gas projects

Russia is strongly integrated into the global organization of production as a leading producer and exporter of oil and gas and has been increasing production levels, modernizing the sector, and diversifying export routes. This expansion has occurred despite the new realities of the global energy agenda that imply transitioning to a more reliable, affordable, low-carbon, and sustainable supply of energy (Bradshaw, 2013). Energy transition policies are rapidly emerging in countries that import oil and gas from Russia

(Khrushcheva & Maltby, 2016). Due to this and other economic and geopolitical reasons, the demand for Russia's fossil fuels may decline, thus affecting Russia's energy security and sociopolitical stability, which currently rely on hydrocarbon export (Aalto, 2011).

To maintain and increase oil and gas production levels, companies have been looking into unconventional extraction methods (e.g., heavy oil extraction in the Yarega field). Moreover, extensive and complex networks of new generation energy infrastructure have been being developed in resource frontiers—very remote regions of Eastern Siberia and the Far East previously of little interest due to their remoteness, harsh climate conditions, and high levels of required investments. New extraction and transportation projects enabled increased volumes of oil and gas to be exported to China and other countries in Eastern Asia. The Eastern Siberia-Pacific Ocean (ESPO) oil pipeline was commissioned during the late 2000s and was followed by the development of the Power of Siberia natural gas pipeline system that became operational in late 2019. Both projects are oriented toward China and the broader East Asian market. The sector's modernization also comes with rapidly expanding capacity for the production and export of LNG. Over the last decade, several LNG projects were brought online, including Sakhalin-2 LNG and Yamal LNG. This diversification occurs in the face of economic sanctions and geopolitical challenges to exporting fossil fuels to traditional destinations in Europe. The shifting spatialities of oil and gas in Russia have been accompanied by growing community concerns.

Rising community concerns

Northern regions of Russia are sparsely populated, apart from a few large industrial towns. Communities at the frontline of energy projects are predominantly Indigenous and rural, maintaining subsistence and semi-subsistence livelihoods based on reindeer herding, hunting, fishing, and gathering. In many areas, oil and gas extraction projects have already resulted in irreversible changes to the environment (due to extensive pollution of waters and land, forest clearance and animal disturbance), access to land and opportunity to practice traditional livelihoods, local economies, social and power relations, unique ecological knowledges and cultures, and many other tangible and intangible aspects of living off the land (Wilson & Istomin, 2019). Extensive networks of leaking pipelines crisscross the taiga in the Khanty-Mansi region in Western Siberia, with enclave-like development affecting the access of Indigenous reindeer herders to their lands (Tysiachniouk & Olimpieva, 2019).

A number of studies have documented community concerns related to oil and gas projects in Komi and Yakutia. The northern parts of the Komi Republic have been an area of extensive environmental pollution, with oil spills occurring persistently since the 1990s (a catastrophic oil spill happened in 1994 on the Kolva River) (Walker et al., 2006). Komi communities are concerned with the impacts of spills of oil and produced waters resulting from oil extraction and transportation, including impacts on the local environment as well as human and animal health (Stuvoy, 2011). In Yakutia, the construction of the ESPO and Power of Siberia pipelines, as well as a network of roads and line clearings, crosses the lands used for subsistence activities by the Evenki people in Eastern Siberia. Forest clearance, land disturbance, noise pollution, and impacts on fish and animals including through poaching raised concerns for the Evenki communities. These concerns include disruption to traditional practices of reindeer herding, hunting, and fishing, potentially resulting in a reduction in food supplies and income (Yakovleva, 2011).

Evolving community responses

While there is a rich and growing literature on the public responses to conventional and emerging energy projects (Boudet, 2019), only a few studies have addressed responses that are prevalent in Russia. Some have suggested that the general public in Russia is passive: people have no genuine interest in or understanding of the oil sector (Poussenkova & Overland, 2018). At the same time, civil society engagement with the oil sector in Russia is rich with lively and varied public debates, some of which take place in social media outlets (Poussenkova & Overland, 2018). However, the impact of public debate and civil society on the energy sector is low, as key mass media are controlled by the state-owned gas company, Gazprom, and the government exercises tight control over different aspects of the economy and society, including those on the Internet (Poussenkova & Overland, 2018). Additionally, the growing resemblance with an authoritarian state has further reduced the space for environmental and climate activism led by Greenpeace Russia and a few environmental and Indigenous NGOs (Tysiachniouk, Petrov, et al., 2018).

Few studies have explored the responses of people actually living in northern regions of Russia where oil and gas are being extracted and energy infrastructure is being built. Across northern Russia, community responses to oil and gas extraction and transportation projects have varied, with direct opposition becoming increasingly common across remote regions. In Siberia, protests were held in 2004 in relation to the route of the ESPO, achieving the goal of rerouting the pipeline to avoid significant ecological damage (Yakovleva, 2014). Many protests have been held in northern parts of the Komi Republic (Pierk & Tysiachniouk, 2016). In the protests taking place in 2014–16 in the Izhma and Usinsk districts, people demanded a halt to extraction if urgent actions to replace leaking pipelines were not taken by the industry (Rodriguez & Loginova, 2018). Apart from expressing concerns over the projects' environmental and socioeconomic impacts, communities are demanding more meaningful participation in resource development (Loginova & Wilson, 2020).

Growing demands for meaningful participation

Community participation in resource development has two dimensions: in the decision-making process and in the benefits stemming from resource projects. In Russia, the formal scope for community participation in decision-making is limited to public hearings as part of the environmental impact assessment of the proposed infrastructure. However, over the years of resource extraction, remote northern communities have developed an awareness of advisable international practices for community participation. The investments in some of the energy projects made by multinational corporations and international lenders meant "importing" international and best-practice rules around community participation, changing expectations from the rules prevalent in the Soviet period when infrastructure projects were centrally planned and delivered and in which community participation was not expected (Tysiachniouk, Tulaeva et al., 2018).

For the indigenous population in northern Russia, these changing expectations include the right for free, prior, and informed consent (FPIC), which means that indigenous people must be informed and consulted about large projects prior to the beginning of development on the territories of traditional land use (Buxton & Wilson, 2013). Moreover, communities are seeking to participate in sharing benefits derived by the developers, which entails that communities that grant access to their traditional territories and resources should receive a share of the benefits, including monetary and nonmonetary benefits. In Russia, such benefits are often codified in socioeconomic agreements and corporate social responsibility (CSR) programs (Tysiachniouk, Petrov, et al., 2018). Socioeconomic agreements are negotiated between companies, local and/or regional authorities and may include representatives of Indigenous communities and institutions. CSR programs in general target contributions to municipalities supporting social infrastructure and the environment.

In practice, remote communities have little to no capacity for negotiation, being powerless actors in their attempts to influence the decisions made in the Kremlin or the offices of state-owned corporations and multinational firms. Communities experience frustration, deception, anger, and community division (Loginova & Wilson, 2020). Part of the problem lies in community experiences of exclusion and nonparticipation in the development of these projects.

Strategies of exclusion and nonparticipation

The exercise of extractive-based development is known to undermine participation practices (Schilling-Vacaflor, 2017). If inclusion relates to meaningful participation practices at different stages of the project development, then exclusion refers to the processes that lead to nonparticipation or a lack of participation. Previous studies identified the exclusionary aspects of community participation in resource projects (Mercer-Mapstone et al., 2019). Bebbington et al. (2013) suggested that a range of strategies constitute "the power of exclusion" (Hall et al., 2011), including discursive strategies, market mechanisms, and legal and bureaucratic strategies (Bebbington & Bury, 2013). This section identifies these strategies as they evolved in the case study communities in the Republic of Komi and the Republic of Sakha (Yakutia). In addition, I identify strategies of uncertainty as a powerful element of nonparticipation.

Discursive strategies

As defined by Bebbington et al., discursive strategies of exclusion are centered on the framing of development and the definition of countries and regions as being naturally predisposed for resource extraction (Bebbington & Bury, 2013). Indeed, in Russia, oil and gas projects are celebrated as an imperative driver of economic growth and development. Over the last decade, the oil and gas sector has been generating up to a half of federal budget revenues and contributes up to one-third of gross domestic product.

In northern regions, energy projects are framed as a pathway to modern development in the absence of other productive and profitable industrial alternatives. This framing is conveyed through regional strategies of industrial and socioeconomic development, programs of CSR, as well as regional and local media and billboards placed in remote villages and regional centers.

Another framing of the energy projects in Russia places them in a broader context of the ideological construction of Russia as a great hydrocarbon superpower (Bouzarovski & Bassin, 2011). Large oil and gas projects are imagined and governed as related to geopolitical processes. As indicated in interviews with community members and regional authorities, the national government is pursuing large-scale energy infrastructure to achieve geopolitical goals, unconditionally promoting and supporting fossil fuel projects. For example, the growing geopolitical importance of Russia—Asia relations has been at the center of the government strategy, media reporting, and community perceptions related to the Power of Siberia gas pipeline and the ESPO oil pipelines. Similarly, the conquest of the Arctic has served as a powerful narrative to justify the rapid development of oil and gas exploration and extraction projects in high latitudes, despite large risks, uncertainties, and the fragility of Arctic environments and cultures due to climate change and modernization. Extensive and resilient energy infrastructure in these resource frontiers is of strategic importance to ensure sustained flows of oil and gas for domestic consumption and exports. Regional governments are often placed "in charge" of nationally strategic projects to ensure there are no constraints on developing resource fields and building infrastructures in a timely manner. Note, for instance, the quotation of a representative of the regional administration of the Komi Republic:

> *We create all conditions for effective industrial operations in our region; even if we cannot help, we make sure that nothing constraints these activities. Of course, there are tensions between strategic projects and people in localities, but they have to understand that these projects are of not only regional but national and international significance (interview, June 2015).*

Discursive strategies hinder the participation of local communities, with limited space to express their opinions and visions for development. According to interviews, community members perceive that regional governments strongly support fossil fuel projects and resource companies. The oil and gas industry is expected to form a significant share of the regional economy and job opportunities. For example, in Yakutia, the rapid expansion of the oil industry resulted in 16% of the regional domestic product's annual growth in 2015. Regional authorities are proud of the infrastructure they helped to create, highlighting the importance of partnership relations with the resource companies and the federal government.

Market mechanisms

The second range of strategies refers to the use of market mechanisms. These primarily operate through the power of economic benefits, compensation mechanisms, and CSR programs (Bebbington & Bury, 2013). They are operationalized by companies seeking to gain legitimacy for the resource projects and authority on the territories of traditional land use. In both the Republics of Komi and Sakha, resource corporations have increasingly been providing finances to support socioeconomic development and address environmental impacts of industrial activities under the CSR agenda. These initiatives include contributions to social infrastructure (e.g., renovation of local schools and hospitals), culture and sports (e.g., support to traditional celebrations and sports events), and restoration of biological resources (e.g., introducing fish to polluted rivers) (Tulaeva & Tysiachniouk, 2017). The payments are usually made to municipal authorities or regional (subnational) governments in the Komi Republic and directly to communities in Yakutia (Gavrilyeva et al., 2019).

Across northern Russia, municipal and regional authorities are tasked with local and regional socioeconomic development. However, according to an interview with a municipal head in the Komi Republic, municipalities in remote northern regions have limited sources of income. Thus, any contribution offered by companies is seen as significant; municipalities are placed in a situation of having marginal negotiation power and accept it. In the Komi Republic, socioeconomic agreements are made between resource companies and regional/municipal governments, commonly without the participation of a local community. Only recently, to overcome this imbalance, the movement of Komi-izhma people initiated a negotiation directly with the oil company, resulting in a contribution (educational opportunities for young community members) that would be locally beneficial according to the community members, instead of going to bureaucrats (Loginova & Wilson, 2020).

Some communities directly affected by large-scale infrastructure projects receive one-time compensations for the loss of land and livelihoods. In Yakutia, during the construction of the ESPO pipeline, compensations for the loss of land and livelihoods were provided to a few families based on land rights and negotiations between representatives of families and companies. Community—company—government relations based on power imbalances have influenced the bases that make resource projects possible through the transfer of land from traditional nature use to industrial (Gavrilyeva et al., 2019).

Legal and bureaucratic strategies

The third range of strategies includes the use of legal and bureaucratic mechanisms (Bebbington & Bury, 2013). In northern Russia, these mechanisms refer to the poor adherence to the state legal framework and limited engagement with global standards for FPIC and meaningful participation. Several previous studies have indicated that existing legislation that regulates local communities' participation in the decision-making process and benefit sharing is not sufficient for stakeholders to consider all local concerns and interests (Gavrilyeva et al., 2019; Wilson & Istomin, 2019). According to the regulations of the Environmental Impact Assessment process, community members can provide their feedback on the project design and companies report the results of public hearings for the government to make a decision. However, community engagement is very limited due to existing regulations, prevailing approaches, and a lack of understanding of international practice (Gulakov et al., 2020).

In Komi, interviews with community members showed that public discussions are seen by companies as being a formality. Experiences of Komi communities, as identified through interviews, demonstrate that organizers of public hearings (municipal administration and project proponents) can control who is invited to public meetings, preventing the participation of community members who dissent. Several community members reported cases when the development of projects had begun before the approval of impact assessment was granted (Loginova & Wilson, 2020).

Another dimension of legislative strategies refers to the nonrecognition of Indigenous status or territorial rights of communities affected by resource projects, thus excluding communities from meaningful participation, as advised by global standards. Global standards for FPIC and meaningful participation do not apply in the Komi Republic, as Komi people are not recognized as indigenous according to national legislation (though they are recognized as such internationally). The situation is different in Yakutia, where the regional government has implemented comprehensive frameworks for indigenous rights, use of traditional territories and benefit-sharing agreements (Gavrilyeva et al., 2019). Despite a robust policy framework, on-the-ground experiences of communities affected by large-scale energy infrastructure suggest that a genuine FPIC from the local population is not always received, as the focus is on compensation for the loss of traditional lands.

Strategies of uncertainty

The final group of strategies is associated with community concerns about a lack of comprehensive, accurate, timely, and accessible information about projects and their progress, as well as poor communication and a perception that communication was deliberately exclusionary. This uncertainty relates to both the information about the participation process and also regarding the impacts and benefits that might accrue from the projects. Indeed, not providing or restricting access to information at the community level can be in the interest of corporations, illustrating "everyday" processes of exclusion (Hall et al., 2011).

Interviews with community members in Komi and Yakutia indicate that communities are concerned with the lack of knowledge and sufficient information about technical aspects of project development and their impacts on health and the environment. Several community members reported that there is a lack of communication regarding these topics, which is intentional, from the company perspective. For example, communities in northern Komi reported that there were cases when companies operating regionally were hiding oil spills and did not report forest clearings. In interviews, community members reported these cases as disinformation and violations of formal processes established by federal and regional regulations.

Most importantly, as interviews show in both Komi and Yakutia, communities lack a clear understanding of formal participation procedures and the arrangements for benefit provision. Formal instruments for the assessment of the loss of land and procedures for the provision of compensations have been obscure and nontransparent for community members. For example, community members in northern Komi Republic reported occasions when only those selected by the local administration could attend public meetings dedicated to the environmental impacts of oil extraction. The recent rise in collective action among remote communities in fighting for more meaningful participation, however, shows that power imbalances stemming from the strategies of uncertainty can be addressed. For example, in northern Komi, the results of community-led assessments of environmental damage differed from the ones performed by the companies, empowering communities to demand a change in the way resource projects are being developed and community participation takes place.

Discussion and conclusion

The development of large-scale energy projects should be accompanied by best-practice community engagement practices and meaningful community participation. Lack of participation, poor communication, and distrust increase the likelihood of community resistance to fossil fuel projects (Conde & Le Billon, 2017). I contribute to the understanding of public responses to the development of oil and gas projects in northern Russia by uncovering challenges of community participation in decision-making and the distribution of benefits. Specifically, I propose that more attention should be given to the multiscalar relational dynamics that contribute to the power of exclusion as it relates to global discourses and national priorities. This approach may enrich our understanding of public responses to the fossil fuel export as it uncovers contextual multifaceted relational features directly linked to public attitudes and community action. In turn, these features can explain the emergence of conflicts and provide opportunities for improving relational justice. These aspects were empirically demonstrated by presenting results from a qualitative study of participation experiences across northern Russia, in particular in the Republics of Komi and Sakha (Yakutia).

Shifting geographies of fossil fuel export in Russia brought large-scale infrastructure projects for extraction, processing, and transportation of oil and gas in northern Russia. These projects are linked to significant transformations at the local and regional scales involving land-use change, environmental degradation, economic development, and cultural impacts. In these remote northern communities, meaningful community participation is demanded, despite prevailing norms and rules that community consent can be taken for granted. On-the-ground experiences of community participation in the development of these projects have played and continue to play a defining role in the way communities perceive the risks and benefits of the projects and respond. Across northern Russia, communities experience numerous challenges because of exclusionary aspects of participation.

I identified four kinds of strategies of exclusion that work together in making projects socially feasible. Discursive strategies project the image of Russia as a "hydrocarbon power," implying that oil and gas projects are imagined and managed in relation to strategic and geopolitical processes, and minimizing the need for community consent. Market-based strategies include the economic benefits directed by oil and gas companies to remote communities where other sources of income are limited, providing communities with marginal negotiating power. Legal and bureaucratic strategies target certain groups or individuals to be excluded (e.g., from attending a public hearing or receiving compensations for land and livelihood loss). Finally, strategies of uncertainty are linked to the accessibility and quality of information about the projects and their impacts, lack of transparency of decision-making, and lack of capabilities in communities to make informed decisions. Although community—company relations evolve differently in each locality, an illusory consensus owes more to the lack of community experience and the specific culture of nontransparent, top-down decision-making in Russia. In this context, opportunities for people to participate in decision-making processes meaningfully are compromised, and the space to make an informed decision about projects is minimized. These exclusion strategies are anchored in the multiscalar space constituted by the interactions of multiple actors and their agendas: the profit-oriented agendas of corporations, geopolitical aspirations of the federal government, and economic development goals of regional governments.

In this chapter, I demonstrate that understanding the experiences of community participation is helpful for research on public responses to fossil fuel export projects as it leads to more reflexivity about contextual community experiences. The different arenas of community participation need to be further explored by providing accounts that seek to unveil ways to minimize exclusion and relational injustice in the context of resource development and export.

References

Aalto, P. (2011). The emerging new energy agenda and Russia: Implications for Russia's role as a major supplier to the European union. *Acta Slavica Iaponica, 30*, 1—20.

Bebbington, A., Bebbington, D. H., Hinojosa, L., Burneo, M. L., & Bury, J. T. (2013). Anatomies of conflict: Social mobilization and new political ecologies of the andes. In *Subterranean struggles: New dynamics of mining, oil, and gas in Latin America* (pp. 241–266). University of Texas Press. https://doi.org/10.7560/748620

Bebbington, A., & Bury, J. T. (2013). Subterranean struggles: New dynamics of mining, oil, and gas in Latin America. In *Subterranean struggles: New dynamics of mining, oil, and gas in Latin America* (pp. 1–343). University of Texas Press. https://doi.org/10.7560/748620

Boudet, H. S. (2019). Public perceptions of and responses to new energy technologies. *Nature Energy, 4*(6), 446–455. https://doi.org/10.1038/s41560-019-0399-x

Bouzarovski, S., & Bassin, M. (2011). Energy and identity: Imagining Russia as a hydrocarbon superpower. *Annals of the Association of American Geographers, 101*(4), 783–794. https://doi.org/10.1080/00045608.2011.567942

Bradshaw, M. (2013). *Global energy dilemmas*. Polity.

Buxton, A., & Wilson, E. (2013). *FPIC and the extractive industries: A guide to applying the spirit of free, prior and informed consent in industrial projects*. International Institute for Environment and Development.

Conde, M., & Le Billon, P. (2017). Why do some communities resist mining projects while others do not? *Extractive Industries and Society, 4*(3), 681–697. https://doi.org/10.1016/j.exis.2017.04.009

Gaventa, J. (1982). *Power and powerlessness: Quiescence and rebellion in an Appalachian valley*. University of Illinois Press.

Gavrilyeva, T. N., Yakovleva, N. P., Boyakova, S. I., & Bochoeva, R. I. (2019). *Compensation for impact of industrial projects in Russia to indigenous peoples of the north* (pp. 83–104).

Graybill, J. K. (2017). Nodes, networks and inefficiency: Understanding Russia's energy landscapes. In *Handbook on the geographies of energy* (pp. 280–295). Edward Elgar Publishing Ltd. https://www.elgaronline.com/abstract/edcoll/9781785365614/9781785365614.xml.

Gulakov, I., Vanclay, F., Ignatev, A., & Arts, J. (2020). Challenges in meeting international standards in undertaking social impact assessment in Russia. *Environmental Impact Assessment Review, 83*. https://doi.org/10.1016/j.eiar.2020.106410

Hall, D., Hirsch, P., & Li, T. M. (2011). *Introduction to powers of exclusion: Land dilemmas in Southeast Asia*. University of Hawai'i Press.

International Energy Agency. (2020). *World energy balances: Overview*. https://www.iea.org/reports/world-energy-balances-overview.

Jentoft, S. (2017). Small-scale fisheries within maritime spatial planning: Knowledge integration and power. *Journal of Environmental Policy and Planning, 19*(3), 266–278. https://doi.org/10.1080/1523908X.2017.1304210

Khrushcheva, O., & Maltby, T. (2016). The future of EU-Russia energy relations in the context of decarbonisation. *Geopolitics, 21*(4), 799–830. https://doi.org/10.1080/14650045.2016.1188081

Loginova, J., & Wilson, E. (2020). "Our consent was taken for granted". A relational justice perspective on the participation of Komi people in oil development in northern Russia. In R. L. Johnstone, & A. Merrild (Eds.), *Regulation of extractive industries: Community engagement in the Arctic*. Routledge.

Mercer-Mapstone, L., Rifkin, W., Louis, W., & Moffat, K. (2019). Power, participation, and exclusion through dialogue in the extractive industries: Who gets a seat at the table? *Resources Policy, 61*, 190–199. https://doi.org/10.1016/j.resourpol.2018.11.023

Pierk, S., & Tysiachniouk, M. (2016). Structures of mobilization and resistance: Confronting the oil and gas industries in Russia. *Extractive Industries and Society, 3*(4), 997–1009. https://doi.org/10.1016/j.exis.2016.07.004

Poussenkova, N., & Overland, I. (2018). Russia: Public debate and the petroleum sector. In I. Overland (Ed.), *Public brainpower*. Springer.

Rodriguez, D., & Loginova, J. (2018). Fluid identities and agendas of socio-environmental movements. In E. Apostolopoulou, & J. A. Cortes-Vazquez (Eds.), *The right to nature: Social movements, environmental justice and neoliberal natures*. Routledge-Earhscan.

Schilling-Vacaflor, A. (2017). 'If the company belongs to you, how can you be against it?' Limiting participation and taming dissent in neo-extractivist Bolivia. *Journal of Peasant Studies, 44*(3), 658–676. https://doi.org/10.1080/03066150.2016.1216984

Stuvoy, K. (2011). Human security, oil and people. *Journal of Human Security*, 5–19. https://doi.org/10.3316/JHS0702005

Tulaeva, S., & Tysiachniouk, M. (2017). Benefit-sharing arrangements between oil companies and indigenous people in Russian northern regions. *Sustainability, 9*(8). https://doi.org/10.3390/su9081326

Tysiachniouk, M., & Olimpieva, I. (2019). Caught between traditional ways of life and economic development: Interactions between indigenous peoples and an oil company in numto nature park. *Arctic Review on Law and Politics, 10*, 56–78. https://doi.org/10.23865/arctic.v10.1207

Tysiachniouk, M., Petrov, A. N., Kuklina, V., & Krasnoshtanova, N. (2018). Between Soviet Legacy and corporate social responsibility: Emerging benefit sharing frameworks in the Irkutsk Oil Region, Russia. *Sustainability, 10*(9). https://doi.org/10.3390/su10093334

Tysiachniouk, M., Tulaeva, S., & Henry, L. A. (2018). Civil society under the law 'on foreign agents': NGO strategies and network transformation. *Europe-Asia Studies, 70*(4), 615–637. https://doi.org/10.1080/09668136.2018.1463512

Walker, T. R., Habeck, J. O., Karjalainen, T. P., Virtanen, T., Solovieva, N., Jones, V., Kuhry, P., Ponomarev, V. I., Mikkola, K., Nikula, A., Patova, E., Crittenden, P. D., Young, S. D., & Ingold, T. (2006). Perceived and measured levels of environmental pollution: Interdisciplinary research in the subarctic lowlands of Northeast European Russia. *Ambio, 35*(5), 220–228. https://doi.org/10.1579/06-A-127R.1

Wilson, E., & Istomin, K. (2019). Beads and trinkets? Stakeholder perspectives on benefit-sharing and corporate responsibility in a Russian oil province. *Europe-Asia Studies, 71*(8), 1285–1313. https://doi.org/10.1080/09668136.2019.1641585

Yakovleva, N. (2011). Oil pipeline construction in Eastern Siberia: Implications for indigenous people. *Geoforum, 42*(6), 708–719. https://doi.org/10.1016/j.geoforum.2011.05.005

Yakovleva, N. (2014). Land, oil and indigenous people in the Russian north: A case study of the oil pipeline and Evenki in Aldan. In *Natural resource extraction and indigenous livelihoods: Development challenges in an era of globalization* (pp. 147–178). Ashgate Publishing Ltd. http://www.ashgate.com/isbn/9781409437789.

CHAPTER 10

Indigenous ambivalence? It's not about the pipeline …: Indigenous responses to fossil fuel export projects in Western Canada

Clifford Gordon Atleo (Niis Na'yaa/Kam'ayaam/Chachim'multhnii)[1], Tyla Crowe[1], Tamara Krawchenko[2] and Karena Shaw[3]
[1]School of Resource and Environmental Management, Simon Fraser University, Burnaby, BC, Canada; [2]School of Public Administration, University of Victoria, Victoria, BC, Canada; [3]School of Environmental Studies, University of Victoria, Victoria, BC, Canada

Introduction

When Indigenous peoples in Canada hear that something is in the "national interest," they know that their rights are about to be ignored or infringed upon. Canada was built on the theft of lands and resources, and the near eradication of Indigenous peoples. But two decades into the 21st century, Canada finds itself in the midst of an era of reconciliation and "nation-to-nation" relationships. Additionally, since the early 2000s, many resource corporations have negotiated Impact and Benefit Agreements (IBAs) with First Nation communities. Enlisting Indigenous support for projects has become widely accepted as a new cost of business. Indigenous peoples have, in some cases, taken on more collaborative roles in these previously one-sided projects. As recent opposition to several pipeline projects in North America has reminded us however, there remains a diversity of responses to the building of oil and gas infrastructure. We contend that a consistent underlying foundation for these wide-ranging responses is the inherent right to Indigenous self-determination, as expressed through ongoing battles over jurisdiction and the responsibility to protect people and territories. We develop this argument by focusing on the apparent ambivalence expressed in Indigenous responses to the

Public Responses to Fossil Fuel Export
ISBN 978-0-12-824046-5
https://doi.org/10.1016/B978-0-12-824046-5.00008-4

proposed $12.6 billion Trans Mountain Expansion (TMX) pipeline project in Alberta and British Columbia, which will facilitate oil export from Western Canada to global markets.

Settler colonialism and the ongoing struggles for Indigenous self-determination

Indigenous peoples are diverse. British Columbia (BC) alone is home to 198 First Nations, whose people and territories exhibit tremendous topographic, climatic, cultural, and linguistic diversity. Unlike much of the rest of Turtle Island (North America), the vast majority of BC lands do not fall under any treaty agreements. The phrase "unceded lands" stems from this lack of formal agreements between Indigenous and settler peoples and the reality of settler colonialism in Canada. Patrick Wolfe described settler colonialism as a distinct structure and not simply a historical event, in which colonials focus on resource and land acquisition, and require the "elimination of the native" (Wolfe, 2006). Most Indigenous territories in BC were usurped by the Crown, despite protests, appeals, and sometimes, physical confrontation. Treaty nations in Alberta have similar, albeit slightly different, relationships with the state and resource corporations. The timing and pace of colonialism is also relevant in Canada. While John Cabot sailed from England to Beothuk and Mi'kmaq territories in 1497, European explorers did not reach the west coast of British Columbia by sea until the late 18th century and did not begin settling in earnest until the mid-19th century. The technologies and ravenous appetite for resources were more advanced upon "first contact" with Europeans in BC, and early encounters were marked with intensified trading and resource exploitation.

Despite the technological imbalance, Indigenous peoples were not timid in their early interactions with Europeans. And while they quickly established trading relationships, this did not preclude conflict. In 1803, Chief Maquinna attacked the British vessel *Boston*, after the ship's captain insulted him, burning the ship and killing all but two of the crew (Union of BC Indian Chiefs, 2005). And in 1811, The US ship *Tonquin* was seized and the crew killed after the ship's captain threw furs in the face of the Tla-o-qui-aht chief (Union of BC Indian Chiefs, 2005). Although coastal Indigenous peoples would be outmatched by the military might of the colonials, it was disease that reduced their populations by 65% to 90% that inhibited their ability to resist further incursions (Inglis & Haggarty, 2000). By the mid- to late 19th century, colonialism became permanent with the formation of

Canada, which usurped Indigenous lands. With the *Indian Act*, the federal government assumed control over Indigenous lives, including a ban of cultural and religious ceremonies, sanctioning of residential schools, and prohibiting the right to organize politically or legally. Most of these restrictions were lifted by 1951. Through increasingly diverse means, Indigenous leaders continued to fight for their rights to self-determination. Resolutions were now sought via court cases, political lobbying and negotiations, but a number of conflicts, notably Oka and Burnt Church, reminded us that military/police enforcement remains likely state responses (King, 2014; Ladner & Simpson, 2010).

The Kanien'kehá:ka (Mohawk) people of Kanehsatake sought to stop the expansion of a golf course in 1990 that culminated into a stand-off with Quebec police and the Canadian military known as the "Oka Crisis." 10 years later, in Esgenoôpetitj (Burnt Church), Mi'kmaq lobster fishers came into conflict with local non-Indigenous fishers and Canada's police over a (1760–61) treaty right to earn "a moderate livelihood." Conflicts in BC have persisted over issues ranging from fishing, forestry, and tourism to oil and gas infrastructure. Recently, Indigenous peoples in Canada have adopted explicitly peaceful protest tactics. No movement displayed this more than the #idlenomore protests, which began in 2012 in response to legislative changes that would imperil Indigenous lands and waters—inspiring solidarity demonstrations around the world (The Kino-nda-niimi Collective, 2014). Even though this peaceful posture has been unambiguous, Canada continues to respond with displays of paramilitary force in recent disputes over natural gas infrastructure projects on both coasts—Elsipogtog in 2013 and Wet'suwet'un in 2019–20 (Crosby & Monaghan, 2018).

Political and legal efforts have also shaped the current terrain of resource management. Beginning with the *Calder* decision by the Supreme Court of Canada in 1973, BC has been a hotbed of legal contention over Aboriginal rights. *Delgamuukw* in particular acknowledged the existence of Aboriginal title, with "an inescapable economic component," but also affirmed that Aboriginal rights could be justifiably infringed for the purposes of economic development and the building of infrastructure. Legal victories for Indigenous communities have led to ongoing negotiations about the interpretation and implementation of rights (e.g., *Marshall, Ahousaht et al, Gladstone,* and *Tsilhqot'in*). While Canadian courts refuse to award

Indigenous communities clear-cut victories, they have compelled governments and project proponents to be more proactive with consultation and accommodation of Aboriginal rights, but they have not prevented conflict.

International initiatives have also contributed to the current climate of Indigenous rights and resources in Canada, most notably the United Nations Declaration on the Rights of Indigenous Peoples (UNDRIP). Canada was one of the last countries to remove its objector status, but the government under Prime Minister Justin Trudeau, "has committed to a renewed, nation-to-nation relationship with Indigenous peoples based on recognition of rights, respect, co-operation and partnership, and rooted in the principles of UNDRIP" (Government of Canada, n.d.). Additionally, BC has passed legislation to begin the process of making provincial laws consistent with the UNDRIP. Bill 41 seeks to: "(a) affirm the application of the Declaration to the laws of British Columbia; (b) contribute to the implementation of the Declaration; and (c) support the affirmation of, and develop relationships with, Indigenous governing bodies" Declaration on the Rights of Indigenous Peoples Act (2019, c. 44). This is promising, but there remains considerable debate on whether the UNDRIP principle of "free, prior and informed consent" (FPIC) amounts to an Indigenous "veto" on resource projects. Indigenous experiences suggest that a veto is not likely, yet the concept of consent can hardly be considered credible if the answer must always be yes.

The history of colonialism in Canada has been characterized by consistent and persistent expressions of self-determination in the face of overwhelming efforts to deny it, especially in relation to decisions about Indigenous lands and waters. Importantly, self-determination has been expressed as much through participation in, as resistance to, collective economic activities. Although the latter has been the focus of violent suppression by the state, it is important to note that the former has also frequently been discouraged, if not actively blocked (Lutz, 2008). These dynamics are considered in the context of the TMX Project.

The Trans Mountain Pipeline expansion project

The proposal to expand (by 'twinning') the Trans Mountain Pipeline became public in February 2012, when Texas-based Kinder Morgan received support from oil shippers expand capacity to meet export demand. The existing pipeline, completed in 1953, linked loading facilities in Alberta to refining and distribution points in BC—a roughly 1150-kilometre route. In December 2013, Kinder Morgan initiated an application to triple the capacity of the pipeline to the National Energy Board (NEB), proposing to begin construction in 2017 with the aim of having oil flowing by December 2019.

Reactions to the proposed project were strong and diverse. The government of Alberta was highly supportive, viewing it as essential to the future of the oil industry. Meanwhile, extensive opposition emerged primarily within BC. The intervening years have seen a complex dance of support and opposition to the proposal, expressed through a variety of sites and tactics, including direct action (over 250 arrests); court cases advanced by Indigenous nations, environmental groups, and affected municipalities, and jostling between the two provincial governments. Indigenous concerns about the project have been expressed from its inception (Bellrichard, 2020; Gill, 2019; Paling, 2018; Shaw, 2020).

By 2018, opposition to the project had reached a fever pitch and Kinder Morgan, citing an "unquantifiable risk," threatened to walk away from the project unless certainty was provided. In response, the federal government purchased the existing pipeline for $4.5 billion. The project proponent thus became Trans Mountain, a Crown corporation, although the government emphasized that it did not aspire to maintain ownership. It also strongly hinted that it would like to see Indigenous ownership or part-ownership of the pipeline, sparking several efforts led by Indigenous nations to negotiate a coalition that could purchase it (John, 2019; Purdon & Pallega, 2019; Reuters, 2019). The awkwardness of the federal government's position as both proponent and decision-maker about the pipeline was highlighted when the Federal Court of Appeals judged that the NEB had fallen short in its consideration of the environmental impacts on the marine corridor and its duty to consult with Indigenous groups. While the court ruled that the NEB did not err in its conduct, it did find that "the consultations did not adequately take into account the concerns of Indigenous groups or explore possible accommodation of those concerns" (Government of Canada, 2019) (p. 1). The Government of Canada accepted the Court's findings and reconducted its consultations before declaring on June 18, 2019, that it approved the project for a second time. Although Indigenous and environmental groups again challenged the approval, by mid-2020 all legal routes of appeal were exhausted and the Federal Court of Appeal ruled that the government fulfilled its duty to consult.

As of 2020, Trans Mountain has negotiated IBAs with 58 Indigenous communities along the pipeline route (Trans Mountain, n.d.). IBAs are privately negotiated and legally enforceable agreements that establish formal relationships between Indigenous nations and industry proponents. But consent remains elusive: 53 First Nations in British Columbia (including many the Government of Canada does not consider to be impacted by the proposal) have formed a treaty alliance against the project, and other First Nations have signed a petition against it. Despite this ongoing resistance, construction of the pipeline expansion has begun (Tiny House Warriors – our land is home, 2020; Tuttle, 2020).

Methods

This study employed a case study methodology drawing on academic, gray literature and policy documents alongside descriptive statistics of impacted communities. The cases were selected based on their identification as communities whose "Indigenous interests" would be impacted by the TMX project (Government of Canada, 2019). Indigenous interest has been defined by the Government of Canada as comprising (i) hunting, trapping, and plant gathering, (ii) fishing and harvesting, (iii) other traditional and cultural activities, and (iv) Aboriginal Title and related governance. Based on this assessment, Indigenous groups were assessed as vulnerable to "negligible, minor or moderate impacts" (none were assessed as highly impacted) and assigned a corresponding "depth of consultation" (low, middle or deeper). In total, 92 Indigenous groups were identified by the government as holding "Indigenous interests" in TMX. Of these, the vast majority (80%) of Indigenous groups are located in BC. Almost a third of Indigenous groups (28% out of the total) were identified as requiring the "deepest" level of consultation; of these, the majority are located in BC's Lower Fraser region, near the terminus of the pipeline. Fig. 10.1 indicates First Nation reserves identified by the federal government as having "Indigenous interests" in the project. This only depicts reserve land and not the broader traditional and often unceded territories of First Nations, nor does it show Métis interests or Indigenous communities not formally recognized.

The case study analysis examined how communities responded to the TMX proposal. The literature search included news reports, gray literature, small community papers, First Nations' press releases, and websites. An assessment of each community's position on the project was compiled based on a review of public statements, whether they had signed an IBA and

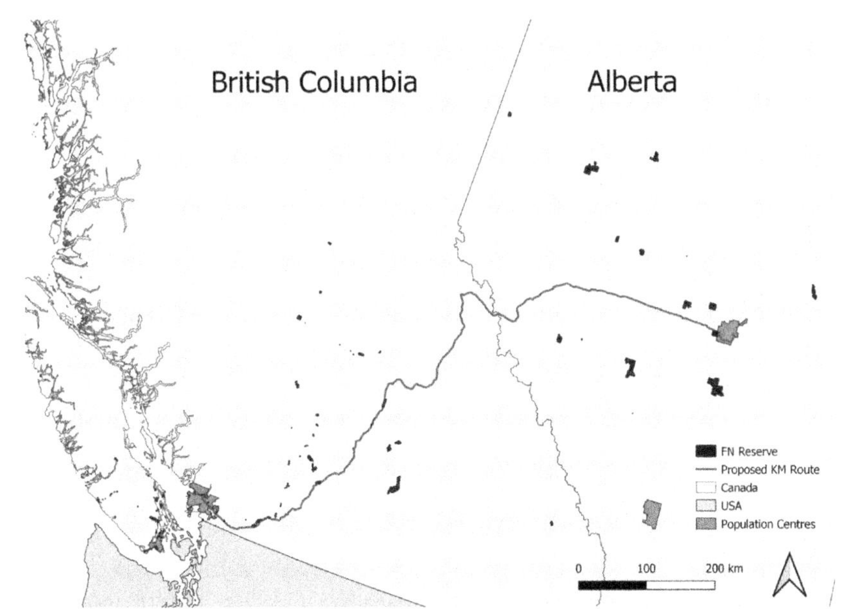

Figure 10.1 National Energy Board "Indigenous Interest" communities along the Trans Mountain Expansion project, 2019 This map depicts all "Indigenous interest communities" as specified in Government of Canada's (2019) Trans Mountain Expansion Crown Consultation and Accommodation Report with the exception of the following: Gunn Métis Local 55 (Lac Ste. Anne Métis, Alberta), Métis Nation of Alberta (Regional Zone 4), Mountain Métis Nation Association, Kelly Lake Métis Settlement Society, British Columbia Métis Federation, Métis Nation of British Columbia. Blue areas (drak gray in print version) on map indicate Census Metropolitan Areas.

where they were in the negotiating process (e.g., engaged, refused to negotiate, and/or had brought court cases against or were otherwise actively opposing the project). The focus was on seeking explicit explanations for the different positions. This analysis was supplemented with an analysis of community characteristics including geography, population, socioeconomic factors, and rights frameworks (e.g., treaty, nontreaty). Descriptive analysis of community characteristics drew on census data, Indigenous Services Canada's Community Wellbeing Index, and First Nation financial reporting. These supplementary data were used to understand community positions within a broader and comparative socioeconomic context.

Case studies: contextualizing ambivalence

The nations along the pipeline route, as well as those consulted due to potential marine impacts, took a range of positions. The continuum of responses stretched from active and consistent support for the project, through support after tough negotiations, through reluctant or "coerced" acceptance, to rejection and ongoing resistance. The characteristics of these nations vary. Around a third are urban communities, a third urban adjacent (within a 1.5-hour drive of a city/town), and the remaining third are rural. They range in population from as small as 13 to over 11,000 (Statistics Canada, 2016). While these communities are diverse, they on average have poorer outcomes across a range of socioeconomic indicators (e.g., income, labor force outcomes) than the non-Indigenous population. For example, individual incomes of First Nation members in our case studies are around half that of the BC and Alberta averages. Despite certain commonalities, historical experiences of colonization have had significant and important differences, depending not least on time of contact and the extent to which and ways in which their lands and resources were appropriated by or entangled with colonial processes (Atleo, 2018). The diversity of responses to TMX should come as no surprise. Understanding how these characteristics and experiences shape those responses presents a more complex challenge.

In this initial scan of cases, we examined public statements from nations about their positions on the proposal. As a consequence of these methods, the cases we focused on are biased toward those who have chosen to take public positions, and as such should not be assumed to be representative of all nations involved. Through these cases, however, we did identify a complex range of responses and rationales; these are captured in Fig. 10.2.

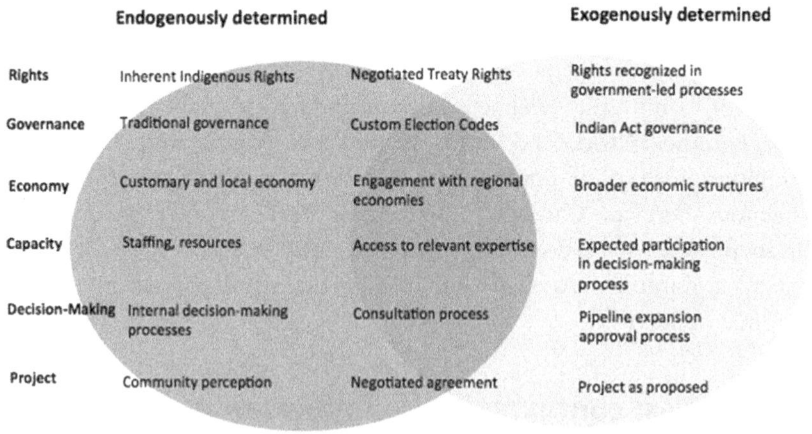

Figure 10.2 Contextualizing Indigenous responses: Key explanatory factors.

Fig. 10.2 identifies the key factors that shape different nations' responses: rights, economy, governance, capacity, decision-making, and project specific. These factors are expressed across a continuum from endogenously determined through exogenously determined and the ways they emerged as significant differs across nations. Below we briefly explain how these factors manifest themselves. Our intention is to illustrate the complexity of the factors that shape nations' responses, and the importance of contextualizing them in ways that respect the diverse conditions that they are navigating. Our underlying argument is that *contextualizing responses to a proposal helps explain why apparently contradictory responses by different nations to a single proposal can nonetheless be understood as consistent expressions of self-determination*. Put more simply, whether the TMX proposal has the potential to support or harm prospects for self-determination, and through what means, varies significantly. We highlight factors that shape that variation. Understanding these factors can help direct research and advocacy attention to better support Indigenous self-determination.

Rights and governance

The top two factors in the figure—rights and governance—have the least direct connection with the responses to the proposal, but consistently surfaced as underlying conditions that shaped community-scale responses. For both factors, the substance of the central "mutually negotiated" category expresses much about the character and strength of the relationships among individual nations, industry, and governments. This in turn shapes several of the factors below: economy, capacity, and decision-making, in particular.

Although there are Constitutional and legal frameworks for the recognition of Aboriginal rights, some nations involved are members of different treaties, while most BC First Nations do not have treaties. Consequently, the scope and character of these rights, and what constitutes appropriate recognition and accommodation of them, varies across cases and is often contested. This creates tensions between how these rights are understood by the nations themselves, and how they are recognized and honored in decision-making processes, especially but not exclusively in BC. For nations in Alberta, more used to living within a treaty framework (although rarely satisfied with the implementation of treaty rights) and more intimately familiar with the oil industry's impacts on the landscape, the scope and scale of rights recognition within these processes is perhaps more clearly defined and less contentious, making this a less salient factor in determining their responses to the proposal.

This spectrum is similarly present in relation to governance, with some nations still living under the 1869 *Indian Act*, including elected councils; others having negotiated "custom election codes," and still others internally juggling two governance systems—hereditary and *Indian Act*—coexisting peacefully or in some tension. Differences around governance processes or rights recognition fundamentally shape the way nations approach a proposal such as the TMX. In turn, their experiences with consultation processes shape their expectations for future engagements.

The interplay of these factors is evident in the number of nations who emphasized their support for the project emerging from a desire to assert jurisdictional authority, and confidence that the project was conducted in an environmentally sound way. As Chief Lampreau of the Simpcw First Nation explained: "We would rather be involved in the process than sitting on the outside and have it pushed through .... At least our concerns are heard right at the front" (Martens, 2019). Similarly, Chief LeBourdais of the Whispering Pines First Nation, an active proponent of the project, stated: "The pipe could send apple juice. I really wouldn't care. I want the environmental oversight and I want the tax authority. That's what I want. For me, it's a jurisdiction issue" (Pimentel, 2018). For these leaders, the project has the potential to enhance their ability to protect their lands and support their communities, thus supporting their self-determination.

The complexities embedded in the judgment over this strategy are apparent in differing responses within the Tk'emlúps te Secwépemc First Nation. Although some members of the nation have consistently been active and vocal opponents of the project (Tiny House Warriors – our land is home, 2020), and the nation initially advanced a court challenge, they ultimately dropped the case when they negotiated an agreement that satisfied the traditional leadership. They stated that an agreement could be a tool used as part of a larger strategy to pursue title to the land ("B.C. First Nations drop out of court challenge, sign deals with Trans Mountain," 2019), and to protect its cultural, spiritual, and historical connections to the land. They emphasized their grounding of this decision in Secwepemc Law: "The potential impacts of the pipeline project on cultural, spiritual, and historical connections to the land were considered, as well as the importance of honoring and upholding Secwepemc law when negotiating the agreement" (Martens, 2019). However, substantial resistance to the project continues to be led by some members of this community.

For the Tsleil-Waututh, their opposition hinges on fundamental infringements of their rights and title. As stated by Chief Leah George–Wilson: "This case is more than a risky pipeline and tanker project - it is a major setback for reconciliation - it reduces consultation to a purely procedural requirement that will be a serious barrier to reconciliation" (Meet the Nations – Pull Together, 2020). Along with their neighboring Squamish First Nation, they advanced legal challenges to the project until all routes were apparently exhausted. They continue to explore alternate routes for opposition (Tsleil-Waututh Nation Sacred Trust, 2015) and express frustration about the consultation process that left many issues unresolved (Bellrichard, 2020).

As the above examples illustrate, issues concerning rights and governance directly shape community responses to the project in diverse ways, and influence many of the factors below.

Economy and capacity

The ways in which economic realities, structures, and concerns shape community responses is similarly complex. Nations have little to no control over the broader economic structure they encounter: capitalism expressed through a regional economy that is often dominated by resource extraction. However, they do have a wide range of involvement with those structures, from participating in wage labor, to revenue sharing or benefit agreements, right through to ownership and control over economic activities. Their degree of involvement has an impact on their capacity, as described below. Nations also vary in the extent to which they rely upon traditional or land-based economies, although there is very little data to assess this.

A nation's economic situation directly shapes its capacity to engage with project proposals in a variety of ways. In all cases the decision of how to proceed is directly affected by the capacity of the nation to understand and evaluate its potential project impacts, benefits, and risks. This in turn depends not only on numbers and expertise of staff but also on the quality of research that is available, and the ability to supplement this by conducting or commissioning its own research. Nations that are well-embedded in the broader economy are more likely to have in-house capacity to respond to proposals, including at one extreme the capacity to conduct their own internal impact assessments, and at the other perhaps no capacity to even access or interpret relevant information. In some cases, capacity can be enhanced by support from other interested parties—such as environmental organizations, the project proponent, or supporters of the industry more widely—simultaneously increasing their capacity to participate in the process but also potentially reinforcing entanglements that might influence their decision.

Some of the strongest opposition to the project has come from those nations—particularly those located in close proximity to Vancouver—that are fairly well-integrated into the broader economy. This affords them greater capacity to mobilize their own research to assess project risks, conduct their own environmental assessment (Tsleil-Waututh Nation, 2015), and advance their legal interests. These nations clearly felt that they had much to lose and had the capacity to resist the project.

Similarly, some of the most vocal supporters of the TMX appear to be entwined with the oil and gas industry, and in some cases the project proponent, in ways that mean the project offers direct benefits to the nation, including strengthening their economies. The Simpcw nation, for example, has worked with Trans Mountain/Kinder Morgan since 2008 through the nation-owned Simpcw Resources Group, whose work includes monitoring and maintaining the existing Trans Mountain Pipeline (Barrera, 2018). The Trans Mountain website discloses that the Simpcw Mutual Benefit Agreement includes milestone and annual payments, commitments of environmental protection, employment, economic and training opportunities, and an emergency response strategy (Trans Mountain, 2018). Likewise, the Alexis Nakota Sioux Nation is already heavily involved in both maintenance of the existing Trans Mountain pipeline and construction of the expansion. Chief Tony Alexis emphasizes that the community is already benefiting from their work with TMX, and he is actively advancing a proposal to purchase the pipeline (Trans Mountain, 2020). Other nations have similarly emphasized the economic benefits they expect the project will bring. Chief Ernie Crey of Cheam First Nation has been an outspoken supporter of the project, emphasizing that their IBA—including a cash component, procurement opportunities, partnerships with companies, training and jobs, environmental protections and an oil spill response plan—will be leveraged as a springboard to other opportunities (Little, 2018). As a form of private contract law, IBAs are generally confidential and their negotiation processes can entail major power/resource and information asymmetries. The confidential nature of agreements means that First Nations cannot share their experiences and collectively bargain with industry proponents in most cases, leaving them at a disadvantage.

For other nations, however, the way economic factors shaped their response to the proposal was not as positive. Chief Ken Hansen of the Yale First Nation expressed that he would not have signed the agreement if his band had any other financial options, but they were "broke" (Paling, 2018). A portion of the IBA money was used to hire staff that could help his administration, which has led to better housing and the hiring of professionals to support the Nation's climb out of debt. As he stated: "Our people needed help and this is one way of getting it." The poignancy of this situation is expressed by April Thomas of Canim Lake, who disagrees with her band's decision to sign an IBA for the pipeline, seeing the pipeline project as "just another divide and conquer tactic that's been used on our people over and over again - the government of Canada made our people so desperate - we have a housing crisis, a poverty crisis and they've made our people so desperate that they feel like they're obligated to sign these agreements because they think that's all they're going to get" (Pimentel, 2018).

The underlying economic constraints were also evident in cases where a lack of capacity to adequately assess the proposal was emphasized. Chief Arthur Bird of the Paul First Nation explained that although they consulted with Elders before signing the IBA, they felt there was not a lot of choice. "We don't have the resources to go and cross examine them and say, 'no, you're wrong.' We don't have the engineers or the biologists to prove anybody wrong. We have to trust the people that are working for us in their dialogue with industry" (Morin, 2017). Similarly, others emphasized their concerns about needing to sign an agreement because they lack capacity to respond to the potential dangers of an oil spill without the resources the agreement would provide. Chief Joseph of the Ditidaht First Nation stated his community does not want the pipeline built because of increased tanker traffic, but they signed the agreement because they believed the project was inevitable. They wanted it in writing that the company would clean up an oil spill: "We agreed not to oppose the pipeline because our first priority has to be to protect our resources, and if we oppose the pipeline we are sitting ducks - fighting is futile" (Beaumont, 2018).

This latter view—that fighting the pipeline was futile—arose repeatedly. The seeming inevitability of the pipeline being built pushed many nations into signing agreements in order to ensure they would receive some benefits to offset both the impacts and the risks of the project. As Chief Joseph stated: "They will not listen to anybody and that's the history of consultation with First Nations people … they consult and go ahead and do what they were going to do anyways" (Beaumont, 2018). This illustrates how the factors above—rights and governance—intersect with those of economy and capacity to encourage nations to effectively negotiate the best deal possible in hopes of improving the resilience of their communities. These factors in turn weighed heavily on decision-making processes.

Decision-making and project

Perhaps the most crucial overarching characteristic of decision-making is that the pipeline expansion approval is entirely exogenously determined: the consultation framework and final decision have always remained with the federal government. There is, however, modest scope for negotiation in the consultation process: nations can argue, for instance, that the assessment of the level of impact was inappropriate and request a deeper level of consultation. They can bring their concerns forward in negotiations over a potential agreement, opening space for mutually engaged decision-making. In turn, all nations control their own internal decision-making, and have chosen a variety of ways to make their decisions: some have consulted with their members, either formally or informally; some have left the decision to leaders, either elected or hereditary; some have held referenda on the nation's involvement (Barrera, 2018; Markusoff, 2018).

As with the decision-making process, the TMX project itself was exogenously determined: nations had no control over their location on the pipeline or tanker route, and were effectively forced to engage with it or to let it proceed without their involvement. Endogenous perceptions of the project were no doubt shaped by many of the factors discussed above, as well as perceptions of how it potentially impacts their lands, but there is no way in which the project itself was shaped endogenously. Some nations, however, felt able to use the consultation and negotiation process to influence the project in ways that made it acceptable.

For some, both the project itself and the decision-making process were deeply problematic, but not in the same way for all. For many, the overarching challenge was the extent to which the final decision had been predetermined. On April 15, 2018, for example (prior to the project receiving approval), Prime Minister Trudeau stated: "We are going to get the pipeline built. It is a project in the national interest … This project will go ahead" (Snyder, 2018). As Chief Aaron Sam Sumexheltza of Lower Nicola, stated: "I could characterize a lot of people as not big supporters of pipelines, generally. But they felt that the project was going to go ahead, and if it was to go ahead, they wanted our community to see the benefits" (Markusoff, 2018). As stated, this sense of inevitability profoundly shaped many nations' decision-making processes.

As described above, some nations had their concerns accommodated through the consultation and negotiation processes, while others continue to oppose the project because of what they see as the failure of the decision-making process to adequately respond to their concerns. The Coldwater Indian Band (CWIB), for example, remains opposed to the project because of the threat it poses to their drinking water. The existing pipeline runs through their reserve and the expansion proposal is planned to be built just outside the reserve boundaries. Although Trans Mountain has promised to work with CWIB, a plan for a new route is still not agreed upon by both parties (Gill, 2019). The perception of being failed by the process also extends to the federal government's involvement: CWIB accuses the federal government of not stepping in when TMX refused to consider alternative routes (Gill, 2019). Similarly, the Sumas First Nation are concerned about the protection of cultural sites and aquifers—a concern accentuated a spill from the existing pipeline on their territory (Barrera, 2020). Despite federal approval for the project being conditional on TMX addressing this concern, they have not yet met this condition. Meanwhile Sumas is filing a specific land claim over the area in hopes of protecting it (Kung & Bissonnette, 2020). Concerns such as these are characteristic of most of the nations who still actively oppose the project: they feel the decision-making process has fundamentally failed them.

Interestingly, despite strong differences of opinion between nations about the project, leaders have generally refrained from criticizing positions taken by others. They have frequently done the opposite, calling for more Indigenous voices to enter into the public debate, and reinforcing a stance that their approval or disapproval should not be used to diminish or counter the position of other nations. Chief Sumexheltza from Lower Nicola Band has stated that signing an agreement does not mean the company has consent to cross the Nlaka'pamux Nation. He supports CWIB's legal opposition to the project. He has emphasized in public statements that nations must have the ability to express their own concerns and make their own decisions: "If you look at First Nations or Indigenous peoples between Vancouver Island and Alberta, it reflects many different communities, many different nations and a diverse set of concerns and opinions on the project.

And I believe it's really important that Indigenous leaders … regardless of what their opinion is, make sure they're vocal …. In many instances, the Indigenous voice hasn't been heard enough. I believe for the benefit of the whole country that everyone's voices be heard" (Stueck, 2018). Indigenous leaders' support for the self-determination of Indigenous nations appears universal, even in a case such as this, where they find themselves on opposite sides of a proposed project.

Although the project was exogenously shaped, some nations either found it acceptable, were able to negotiate minor changes, or negotiated adequate compensation or protection from risks to find it acceptable. For others, their concerns about the project have either never been adequately addressed, or are quite simply too fundamental to the nature of the project to be addressed. Crucially, even among those nations who do approve of the project, and who have stated they are satisfied with their agreements, none have approved of the overall decision-making process, and many frustrations remain—both from those who support and those who oppose. Central among these is the overarching impression that the project would proceed regardless of Indigenous communities' concerns, thus circumscribing the scope of Indigenous self-determination in familiar and problematic ways.

Discussion and conclusions

When confronted with a project such as the TMX, First Nations are between a rock and a hard place. There is a government process in place that aims to recognize that First Nations along the route will be impacted in various ways by the pipeline (though narrowly defined), and specifies levels of consultation to address these impacts. This process, however, does not entail free prior and informed consent as defined by the UNDRIP: industry directly "consults" and negotiates with First Nations amidst vast power, resource, and information asymmetries. The giant shadow over these interactions is that First Nations cannot say "no."

The strategies left to individual First Nations in Canada are to assert their rights to self-determination as they have done throughout colonization: within and against these constraints. It is in this context that despite the variety of Indigenous responses, all can be understood as expressions of self-determination: each nation has made a judgment about under what conditions the project will support their community, and due to the varieties in their situations these judgments are varied. For some nations navigating the long game of negotiating, the project garners their support because it has the potential to strengthen their capacity for self-determination. For others, this and other projects represent existential threats to their nationhood. Although the latter nations in this case feel the constraints of the colonial relationship most acutely, none would deny the constraints.

Indigenous peoples in Canada are not alone in these experiences. There is a large and growing literature on Indigenous responses to extractive industries and long-standing debates regarding the extent to which they adequately include and benefit Indigenous peoples (Altman, 2004; Altman & Martin, 2009; Cameron & Levitan, 2014; Peterson St-Laurent & Billon, 2015). In the case of resource developments, the literature asks how Indigenous peoples can have more collective control over developments on their territories and how communities can avoid the "resource curse"—with short-term economic gains followed by decline and in many cases, environmental degradation (Eggert, 2001; Morgera, 2019; Parlee, 2015; Petrov & Tysiachniouk, 2019). Much depends on how Indigenous rights are recognized in national law—these rights frameworks are often inadequate, as Loginova illustrates in Chapter 5 on the rights of the Sami people in energy infrastructure projects in northern Russia.

Absent meaningful rights recognition and sovereignty over their territories, what strategies might First Nations adopt to strengthen their positions within this context? Among *endogenous* factors, ceteris paribus, what options do First Nations have? The government-prescribed consultation process sets up industry to negotiate directly with individual nations. One strategy is to increase First Nations' capacities to engage in this process. Nations should be adequately resourced to come to the table as informed and empowered participants in this process through access to the relevant expertise: they should not bear these costs. The government of Alberta's Indigenous Opportunities Corporation offers nonrepayable "Capacity Grants" for this purpose, as it is very much in their interests; BC does not. Beyond this, it is important to note that the confidential nature of the IBA process means that First Nations are not able to share leading practices or jointly negotiate. As confidential documents, it is very challenging to ascertain if IBAs are in fact benefitting First Nations at all and there is little in the way of comparative research on these agreements. Broader institutional capacity across nations is important and efforts could be made to relax at least some of the confidentiality provisos to facilitate this. Other options are to organize and advocate dissent to the pipeline, as some nations continue to do. As we have highlighted, however, policed/militarized responses to Indigenous dissent make this a very challenging option.

Where does real change lie? In all those exogenously determined factors that constrain Indigenous sovereignty. It is essential that Indigenous rights are recognized in developments on traditional territories, and that an answer other than "yes" is a possible outcome. As stated by Khelsilem, an elected leader of the Squamish nation: "We have a right to practice our culture, our way of life, and to continue our right to self-determination in our territories. This is a right that we have never surrendered, and it is a right we will continue to defend" (Ritchie, 2018). Understanding Indigenous ambivalence to pipeline proposals as an expression of self-determination helps to refocus attention on the underlying challenges that must be confronted. Indigenous ambivalence, in other words, is not itself the problem, but it does point to the work that will be needed to reconcile self-determination with large-scale infrastructure development under the guise of "national interest."

References

Altman, J. C. (2004). Economic development and indigenous Australia: Contestations over property, institutions and ideology. *The Australian Journal of Agricultural and Resource Economics, 48*(3), 513–534. https://doi.org/10.1111/j.1467-8489.2004.00253.x

Altman, J., & Martin, D. (2009). *Power, culture, economy: Indigenous Australians and mining.*

Atleo, C. (2018). *Change and continuity in the political economy of the Ahousaht.*

Barrera, J. (2018). *B.C. First Nation eyes renegotiating benefit deal on Trans Mountain pipeline project.* CBC.

Barrera, J. (2020). *Stó:lō First Nation eyes claim over lightning rock site in path of Trans Mountain.* CBC.

B.C. First Nations drop out of court challenge, sign deals with Trans Mountain. (2019). National Post.

Beaumont, H. (2018). *Why First Nation chiefs sign Trans Mountain pipeline deals.* Vice News.

Bellrichard, C. (2020). *B.C. First Nations disappointed Supreme Court won't hear their appeal of Trans Mountain project.* CBC News.

Cameron, E., & Levitan, T. (2014). Impact and benefit agreements and the neoliberalization of resource governance and indigenous - state relations in Northern Canada. *Studies in Political Economy, 93*, 25–52. https://doi.org/10.1080/19187033.2014.11674963

Crosby, A., & Monaghan, J. (Eds.). (2018). *Policing indigenous movements e Dissent and the security state.* Fernwood Publishing.

Government of British Columbia. (2019). Declaration on the Rights of Indigenous Peoples. *Bill, 41.* https://www.leg.bc.ca/parliamentary-business/legislation-debates-proceedings/41st-parliament/4th-session/bills/third-reading/gov41-3.

Eggert, R. (2001). *Mining, minerals and sustainable development.*

First nations "extremely disappointed" by supreme court of canada's refusal to hear tmx appeal, but vow to keep fighting, (2020).

Gill, I. (2019). *Why trudeau's Trans Mountain dreams may trickle out in coldwater.* The Tyee.
Government of Canada. (2019). *Trans Mountain expansion crown consultation and accommodation report.*
Government of Canada. (n.d.). *United nations declaration on the rights of indigenous peoples - indigenous and northern affairs Canada.* https://www.afn.ca/uploads/files/education2/undripcanadiangovernments.pdf.
Inglis, R., & Haggarty, J. (2000). Huu-ay-aht First Nations, Kiix?in agenda paper. In A. Hoover (Ed.), *Nuu-chah-nulth voices, histories, objects & journeys* (Vol. 39). Royal British Columbia Museum.
John, G. (2019). *Swell of support from indigenous groups gets Trans Mountain pipeline back on track - energy news for the Canadian oil & gas industry.* EnergyNow.Ca.
King, S. J. (2014). *Fishing in contested waters : Place and community in Burnt Church/Esgenoôpetitj.* University of Toronto Press.
Kung, E., & Bissonnette, M. (2020). *The Trans Mountain pipeline and specific claims.* West Coast Environmental Law.
Ladner, K. L., & Simpson, L. (Eds.). (2010). *This is an honour song: Twenty years since the blockades, an anthology of writing on the "Oka crisis.".* Arbeiter Ring Pub.
Little, S. (2018). *This B.C. First Nation wants to buy a piece of the Trans Mountain pipeline.* Global News.
Lutz, J. S. (2008). *Makúk: A new history of aboriginal-white relations.* John Sutton Lutz. UBC Press.
Markusoff, J. (2018). *Trans Mountain and First Nations along the pipeline route: It's not a dichotomy of "for" or "against" - Macleans.ca. Macleans* Accessed 3 sept. 2021 from https://www.macleans.ca/news/canada/trans-mountain-politics-and-first-nations/.
Martens, K. (2019). *B.C. First Nation reverses decision, signs up for piece of Trans Mountain expansion project.* APTN National News.
Meet the Nations — Pull Together. (2020). *Pull together* Accessed 3 Sept. 2021 from https://pull-together.ca/about-copy/.
Morgera, E. (2019). Under the radar: The role of fair and equitable benefit-sharing in protecting and realising human rights connected to natural resources. *International Journal of Human Rights, 23*(7), 1098–1139. https://doi.org/10.1080/13642987.2019.1592161
Morin, B. (2017). *B.C. grand chief responds to Alberta premier on Trans Mountain, warning it "will never see the light of day".* CBC.
Paling, E. (2018). *B.C. chiefs say they don't support Trans Mountain pipeline despite signing agreements.* Huffington Post Canada.
Parlee, B. L. (2015). Avoiding the resource curse: Indigenous communities and Canada's oil sands. *World Development, 74,* 425–436. https://doi.org/10.1016/j.worlddev.2015.03.004
Peterson St-Laurent, G., & Billon, P. L. (2015). Staking claims and shaking hands: Impact and benefit agreements as a technology of government in the mining sector. *Extractive Industries and Society, 2,* 590–602. https://doi.org/10.1016/j.exis.2015.06.001
Petrov, A., & Tysiachniouk, M. (2019). Benefit sharing in the Arctic: A systematic view. *Resources, 8*(3), 155. https://doi.org/10.3390/resources8030155
Pimentel, T. (2018). *Traveling the pipeline: Why the Secwepemc nation is crucial for the Trans Mountain pipeline.* APTN National News.
Purdon, N., & Pallega, L. (2019). *Trans Mountain pipeline: Why some First Nations want to stop it — and others want to own it.* CBS News.
Reuters, T. (2019). *Indigenous-led group promises $6.9B TMX pipeline bid within week.* CBS News.
Ritchie, H. (2018). *Squamish nation reacts to federal government kinder morgan purchase.* Squamish Chief.

Shaw, K. (2020). Flashpoints of possibility: What resistance reveals about pathways towards energy transition. In *Regime of obstruction: How corporate power blocks energy democracy*. Athabasca University Press.

Snyder, J. (2018). We are going to get the pipeline built. In *Trudeau begins federal talks with Kinder Morgan to guarantee Trans Mountain*. National Post.

Statistics Canada. (2016). *Census of the population, population estimates, July 1, by census subdivision, 2016 boundaries*. Catalogue no. 98-316-X2016001.

Stueck, W., (2018). Amid vocal Indigenous opposition, chief ernie crey speaks out in favour of trans mountain pipeline. The Globe and Mail, accessed 3 Sept. 2021 from https://www.theglobeandmail.com/canada/british-columbia/article-amid-vocal-indigenous-opposition-chief-ernie-crey-speaks-out-in/.

The Kino-nda-niimi Collective. (2014). In *The winter we danced: Voices from the past, the future, and the idle no more movement*. ARP Books.

Tiny House Warriors – our land is home. (2020).

Trans Mountain. (2018). *Simpcw resources group's pipeline contracting work balances jobs and respect for the environment*.

Trans Mountain. (2020). *Meet backwoods energy Services*.

Trans Mountain. (n.d.). *Indigenous benefits - Trans Mountain*.

Tsleil-Waututh Nation. (2015). Assessment of the Trans Mountain pipeline and tanker expansion proposal. In *Sacred trust initiative - tsleil-waututh nation*.

Tuttle, R. (2020). *Arrests escalate protest over Trans Mountain pipeline*. Calgary Herald.

Union of BC Indian Chiefs. (2005). *Stolen lands, broken promises: Researching the Indian land question in British Columbia* (2nd ed.).

United Nations. (2011). United Nations Declaration on the Rights of Indigenous Peoples, A/RES/61/295, accessed 3 Sept. 2021 from https://www.un.org/development/desa/indigenouspeoples/wp-content/uploads/sites/19/2018/11/UNDRIP_E_web.pdf.

Wolfe, P. (2006). Settler colonialism and the elimination of the native. *Journal of Genocide Research, 8*(4), 387–409.

CHAPTER 11

The primacy of place: a community's response to a proposed liquefied natural gas export facility*

Emily Paige Bishop and Karena Shaw
School of Environmental Studies, University of Victoria, Victoria, BC, Canada

Introduction

Over the past decade, the Province of British Columbia, Canada, has aggressively sought to expand the production of shale gas in the Northeast of the province, catalyzing a liquefied natural gas (LNG) export industry on the west coast. The former Premier of British Columbia, Christy Clark, proposed the LNG industry as a "win-win" solution: it would simultaneously establish a new extractive industry within Northern BC to support rural economic development amidst the failing forestry and fisheries industries, and advance a climate change solution in the form of "clean LNG" to replace dirtier coal in Asian export markets. These promises faced resistance from the outset, with critics contesting these claims and highlighting the multiscalar environmental impacts of the industry, the unequal distribution of risks and rewards, and potential impacts to Indigenous rights and title (Garvie & Shaw, 2016; Gilchrist, 2017; Horne & MacNab, 2014; Stephenson et al., 2012).

One of the focal points for opposition was the Pacific Northwest LNG (PNW LNG) terminal, an $11-billion LNG conversion plant and export terminal proposed in 2013 by the Malaysian state-owned energy company Petronas and its partners. If built, PNW LNG would have been the largest

* The research for this chapter was supported by the Corporate Mapping Project, a research and public engagement initiative investigating the power of the fossil fuel industry in Western Canada. The CMP is jointly led by the University of Victoria, Canadian Centre for Policy Alternatives and the Parkland Institute. We would also like to thank Tamara Krawchenko, Eleanor Stephenson, two anonymous reviewers and the editors of this volume for their helpful engagement with earlier drafts of this chapter.

Public Responses to Fossil Fuel Export
ISBN 978-0-12-824046-5
https://doi.org/10.1016/B978-0-12-824046-5.00001-1

single source emitter of greenhouse gas emissions in Canada (Pembina Institute, 2016). Within the project's host city of Prince Rupert—and across the Province—the project was very divisive (Kelly & Morgan, 2016). Proponents welcomed its potential economic benefits, including access to jobs and training (Lough, 2017). The resistance was dynamic and multifaceted, involving numerous groups—from local concerned citizens, grassroots environmental groups, fisher-people, and First Nations to scientists, politicians, and provincial ENGOs (Garvie & Shaw, 2016; Gilchrist, 2017; Horne & MacNab, 2014; Stephenson et al., 2012). Then-Premier Christy Clark dismissed resistance to the PNW LNG proposal, calling opponents "a ragtag group of people" and framing opposition as the "forces of no." In her words: *"I'm not sure what science the forces of 'no' bring together up there except that it's not really about the science. It's not really about the fish. It's just about trying to say no. It's about fear of change. It's about a fear of the future"* (quoted in Gill, 2016). While Clark's characterization of the resistance reflected her government's aggressive push to develop the project—and a familiar effort to delegitimize the arguments of opponents of industrial development (McClymont & O'Hare, 2008: 322)—her dismissal played on a limited understanding of community-scale responses to energy development more broadly.

In this chapter, we develop a grounded case study of the PNW LNG project to draw out key factors driving community responses and resistance. Drawing on field research and interviews with community members and representatives of Environmental Non-Governmental Organizations (ENGOs) who opposed the Pacific NorthWest LNG project, the chapter situates resistance to the project within the trajectory of extractive development in the region. Our findings suggest that, although ENGOs and think-tanks foregrounded climate change mitigation as a primary reason to oppose the proposal, regional resistance to the project was grounded in a deep commitment to place and community, as well as a desire to have a voice in shaping the future of the region. The proposed PNW LNG project exposed a fundamental failure of decision-making processes to engage and respond to community concerns and priorities, within a historical context characterized by unsustainable environmental governance, colonialism, and extractivism.

These findings reinforce the urgency of examining community-scale responses to energy projects amidst both rising pressure to develop new extractive industries and multiscalar resistance to such projects. The intersection of pushes for development and local resistance will shape the future of extractive industries, communities, and environments (Arsel et al., 2016; Keeling & Sandlos, 2016; Veltmeyer & Bowles, 2014). This case exemplifies many of the concomitant uncertainties posed by development and resistance: despite receiving full federal and provincial government approval, in the face of sustained and substantial opposition, Petronas canceled the project in July 2017, citing "market conditions" (Petronas, 2017), although ongoing resistance no doubt also played a role (Cattaneo & Morgan, 2017). The failures of the PNW LNG project offer broader insights into what is needed to support rural communities through transformative change in turbulent times. This is especially critical given the need for rapid energy and industrial transitions to mitigate climate change. Simultaneously transitioning energy systems away from reliance on fossil fuels and reshaping extractive economies requires attentiveness to how these changes are *received by and impact specific communities*. Developing solutions that support communities through these interwoven transitions, and ensuring that the economies that emerge will meet their needs, requires better understanding local values and priorities.

Context

Biersack (2006) speaks to how global forces drastically shape local places, and how even subsistence communities do not exist in equilibrium or isolation, but rather are part of a world system that is impacted by markets, social inequalities, and political conflicts: *"Place is ... the grounded site of local-global articulation and interaction"* (p. 16). The attempt to introduce an LNG industry in Prince Rupert must be understood as nested within this intersection of global and local forces. British Columbia has been shaped by extractive industries and conflicts over them (Barnes & Hayter, 1997; Blomley, 1984; Cashore, 2014; Magnusson & Shaw, 2003; Wilson, 1998). For the past 150 years, Northern BC in particular has depended on resource and extraction-based industries—including mining, fishing, and logging—with economic growth and fiscal revenues made possible by the dispossession of Indigenous peoples from their lands (Harris, 2002); since the first contact with European peoples, *"the North Coast has been a cauldron of extractive industries"* (Menzies, 2015: 8). The centuries-long colonial project has resulted in Indigenous peoples being systematically excluded from access to the economy, from meaningful decision-making about or access to their lands, and thus from the capacity for communities to thrive (Lutz, 2008; Menzies & Butler, 2008).

Although rich in resources, rural communities in BC have witnessed the bulk of this resource wealth leave their communities to support development elsewhere, from colonial accumulation through present development of urban centers such as Vancouver. All three of the industries that initially shaped northern BC—the fur trade, fishing, and forestry—have hit hard times. The fur trade, of course, collapsed fairly early on. Forestry—a consistent employer in the region—has struggled with "fall-down" as a consequence of overharvesting, and was dealt a nearly fatal blow with the globalization of the forest industry in the latter part of the last century, resulting in mill closures in the early 2000s. Poor management of fishing stocks has led to profound challenges in the fishing industry, resulting in stock collapses, closures of canneries, and, more recently, threats to commercial and recreational salmon fisheries (Price et al., 2008), though, as we document, salmon fisheries continue to be both culturally and economically vital on the Northwest coast.

While the specific contours of resource dependency, and its impacts, are unique in different places, the loss or decline of each of these industries has had profound ripple effects across Northern communities: the population of Prince Rupert decreased by nearly 30% with the closure of the Skeena Cellulose pulp mill and the region's last fish canning plants in the early 2000s (Allen, 2018). More broadly, extraction can result in ecological degradation (Garvie et al., 2014; Garvie & Shaw, 2016; Spice, 2018), impacts to clean water (Parfitt, 2017), risks to public health (Hughes, 2017), and various socioeconomic impacts on local communities (Garvie & Shaw, 2014; Stokes et al., 2019).

The decline of the region's traditional staples industries and subsequent economic impacts partially motivated the strong government push to establish a new shale gas industry in northeast BC, with the corresponding LNG liquefaction and export industry in northwest BC. The former Premier of BC, Christy Clark, called LNG an economic opportunity that could transform the province (Renshaw, 2014). Campaigning in the 2013 provincial election, Clarke had stated that an LNG sector would pay off the provincial debt and create more than 100,000 jobs (Zussman, 2016). The quality of the shale gas resource, promise of market demand in Asia, and extensive government support led to a plethora of proposals for pipelines, liquefaction facilities, and export terminals; by 2014 there were at least 18 different proposals vying for approval in the province (Northwest Institute, 2019). These proposals emerged amidst strong, region-wide resistance to another proposed fossil fuel project, the Northern Gateway bitumen pipeline from the Alberta oil sands (Bowles & MacPhail, 2017; Bowles & Veltmeyer, 2014). Some supported the projects because the need for economic development in the region was paramount, and LNG infrastructure seemed a less risky proposition than a bitumen pipeline (Smith, 2019). For others, the momentum that had built in opposition to a bitumen pipeline extended to all proposed fossil fuel export projects.

Against this backdrop, the Pacific Northwest LNG (PNW LNG) proposal emerged as one of the most high-profile and divisive proposals. PNW LNG proposed to construct and operate an LNG conversion plant and export terminal on the northwest coast, as well as the Prince Rupert Gas Transmission (PRGT) pipeline to connect the shale gas fields in northeast BC to the terminal. The plant itself was to be located at the mouth of the Skeena River on the territory of the Tsimshian Peoples, near the community of Prince Rupert, while the PRGT pipeline would span 900 km through First Nations' territories across northern British Columbia (BC Oil and Gas Commission, 2020). If built, the PNW LNG terminal would have shipped 19 million tonnes of liquefied gas annually on tankers destined for markets in Asia (Ghoussoub, 2017).

Resistance to the project was multifaceted and spanned from grassroots groups to national ENGOs. Although affected Indigenous groups were divided on the merits of the project,[1] Indigenous peoples led much of the local resistance, and concerns about Indigenous rights and title were paramount, with a number of First Nations stating that the Pacific NorthWest project *"[did] not meet the test"* for respecting Indigenous rights (de Wolff, & Broten, 2020; Horne, 2015; Jang & McCarthy, 2016). Indigenous resistance to the project focused on Lelu Island, the location of the proposed terminal, where hereditary leaders of the Gitwilgyoots tribe of Lax Kw'alaams, whose territory includes Lelu Island, constructed a permanent settlement and prevented workers from accessing the site location (Horne, 2015). A second land-based resistance effort took place in Gitxsan territory: Camp Madii Lii was constructed on the proposed route of the PRGT pipeline (Jang, 2017). Opposition was supplemented by a range of other tactics, including community engagement and organizing (Friends of Wild Salmon, n.d.; Leadnow, n.d.; Sierra Club of BC, n.d.), multiple legal challenges against the federal government (Linnett, 2016), and the penning of a letter critiquing the Canadian Environmental Assessment of the project signed by 130 Canadian and international scientists (Hume, 2016). Numerous ENGOs in British Columbia campaigned on the project, often

[1] See chapter by Atleo et al. in this volume for richer engagement with how decision-making about projects such as this intersects with Indigenous self-determination.

directing their messaging to broader audiences (Darwish, 2014; Wilderness Committee, 2016), with think-tanks based in southern BC tending to emphasize the project's potential climate impacts (Pembina Institute, 2016).

Methods

To better understand the dynamics of community responses to the PNW LNG proposal, we first reviewed media coverage of the proposal from 2013 to 2017. Then, in the summer of 2018, the first author conducted 3 weeks of field work in the Prince Rupert region, including 18 semistructured interviews with opponents of the project, including community members and representatives of ENGOs. These interviews explored interviewees' perspectives on the proposed project, including what motivated their opposition to it, as well as their visions for the future of their communities. Interviews were coded and analyzed for emergent themes and shared narratives, with particular attention both to convergences and divergences in opponents' perspectives.[2] The interviews produced rich and nuanced conversations; however, our findings are framed by several crucial limitations. Most significantly, as non-local Settler scholars, our analysis is from a distinctly "outsider" perspective. These findings should not be taken as a representation of the full spectrum of affected communities' responses, and the role of colonialism in shaping resistance to such projects, as well as their full implications for Indigenous communities, merit careful consideration that exceeds our scope.

Findings

Interviewees identified a diverse range of concerns inspiring resistance to PNW LNG, but the overwhelming focus was on three prominent issues: (1) the potential localized *environmental impacts*, particularly to salmon runs in the Skeena River; (2) the potential *economic impacts*, particularly the ways that this development would lock the region into another extractivist economy and its attendant economic vulnerability; and (3) the failures of the *governance and decision-making* processes to adequately engage with the concerns raised by local communities.

[2] A more comprehensive analysis of our findings is available in {citation(refId:4e0811f7-8836-4b87-852a-6ffaf5dc173a)}.

"It was about the fish, really; salmon, and the communities that relied on them." (N-13)[3]

The overwhelming issue that drove local opposition to the PNW LNG project was the perceived danger it posed to salmon. These concerns arose because of the proposed location of the terminal on Lelu Island. Lelu Island, which is also known as Lax U'u'la, lies at the mouth of the second largest salmon producing river in Canada, the Skeena River. Adjacent to Lelu Island is Flora Bank, an offshore area rich in eelgrass which provides critical habitat for juvenile salmon from the entire Skeena watershed. As one interviewee explained: "*We focused on Petronas because of where it was located: every stream, every tributary, every single salmon from our entire watershed goes to that one place*" (N-12). The site location posed a threat to vulnerable fisheries, to local subsistence harvesters of salmon, and to the regional economy, to which the Skeena salmon industry contributes $110 million each year (Hume, 2016). Several other LNG export terminals were proposed for the region but did not garner the same levels of resistance; opposition to PNW LNG must be understood in relation to the site location and the impacts that development in that place would have on the region generally, and on salmon in particular. As emphasized by one interviewee: *"If they hadn't tried to put Pacific NorthWest LNG on Lelu there would not have been a problem because the majority of people were for LNG here. So that choice of location made things very divisive"* (PR-5).

Interviewees repeatedly emphasized the threats the project posed to the whole Skeena Watershed, including lands and waters belonging to the Tsimshian, Gitxsan, and Wet'suwet'en Nations, explaining that threats to salmon were threats to the place and the community itself: *"In this part of the world, people are connected to salmon in a major way so if you want people to stand up and care or if you want to get people concerned, threaten their salmon. It was about the fish, really; salmon and the communities that relied on them"* (N-13). Interviewees spoke of salmon as defining the place and what it meant to live there, and as being integral to the ecology, culture, and economy of the region: *"It's the core reason why we're all here, whether we know it or not, whether we interact with it or not, whether we eat it or not, salmon is the reason why people*

[3] Interviews were coded based on location: nine interviewees were based in Prince Rupert (cited as PR 1–9), five were from northwestern BC (indicated as N 10–14), and four were from southern BC (indicated as S 15–18).

are here .... Salmon is the backbone of all survival and industry and everything here. Whether it's symbolic or really personal and physical, it's an undeniable force here" (N-11).

Although this commitment to salmon was the foundation for resistance, interviewees also expressed how deeply salmon have shaped communities: their histories, values, and desires for the future. This keystone species had been under threat prior to the PNW LNG proposal; overfishing, poor fisheries management, and habitat destruction have negatively impacted local fisheries—and the communities that rely on them—for decades, and climate change is anticipated to add to these threats (Walters et al., 2008, p. 115). Communities' concerns of the vulnerability of salmon deeply shaped their responses to this project (see below). However eager communities were for new economic development opportunities, they were not willing to put salmon at risk to facilitate it: *"the lifeblood of BC is not oil and gas, it's salmon"* (S-5).

"You're asking us to take all of the risks—which seem rather significant—and get none of the rewards?" (N-12)

Embedded in local resistance was a desire not just for a different industry to support economic development but for a different kind of economy. While the region's history of boom-bust economic development had left some communities in desperate economic straits—and this economic vulnerability framed the case for LNG—many interviewees questioned this model of economic development. As one interviewee explained, people in the region *"have seen that the boom and bust cycle isn't how they want to live: they don't want to be going from one mega project to the next"* (N-10). For many interviewees, the development of an LNG industry risked reinscribing an old path dependency of overexploiting a resource to stimulate the economy, only to then find the economy threatened as the resource dwindles. One noted that support for LNG has: *"become the same rhetoric that you hear across most of rural BC which is 'jobs, jobs, jobs.' And 'if this industry isn't here, well what will we have?' That dependence and fear-based decision-making that boom-and-bust economies bring"* (N-11). Rather than an economic savior, they argued that an LNG industry risked stimulating yet another—in some ways more extreme—boom-bust economy in the region.

Interviewees also emphasized that they wanted a change in the current model of industrial development that would see benefits from extracting and exporting resources—be it fish, forestry products, or LNG—stay in local communities: *"I'm tired of the fact that more than 72% of our oil sands are owned by China ... that the companies that are investing in LNG were foreign. The benefits, the profits: they don't stay here. The risks: we're the ones taking them. But they get the vast majority of the rewards and I think if we're going to extract our resources, develop them, then I think it should be for our benefit. It should be our decision-making and our control*" (N-12). Deeply frustrated by the perceived historical mismanagement of renewable resources in the region—resources that could and should have sustained the region indefinitely, and that had in the past offered residents good livelihoods—interviewees expressed deep suspicion of a new industry that they could not participate in shaping: "*We, as people who live in the region, have always had a fair degree of control over our lives. We're losing that because we no longer control the resource that's on our doorstep. That's being taken away from us - we no longer control much of our lives, it's all done from somewhere else by someone else. We're not comfortable with that, we'd much rather have an economy that we have a direct hand in and say in, and manage in some fashion*" (PR-1).

While concern about perpetuating boom-bust cycles catalyzed resistance to the project, so did a desire for a different kind of economic future—one that would meaningfully address communities' needs. Interviewees explained that their resistance to the project was not opposition to industry or new economic development outright; they expressed that people are in favor of "*the right kind of industry. And the oil and gas industry is not the right industry*" (PR-9). As stated by another community member, "*it's not all about the resistance for the sake of not building industry. It's about making sure that the growth is sustainable and done in an environmentally responsible way*" (PR-2). Interviewees emphasized that while the PNW LNG proposal had divided communities, "*it also helps force some of this conversation, [this] discussion that we need to have as communities around what we want to see, our future. It caused us to put a bunch of effort into being proactive about defining what we're for*" (N-13). They were interested not just in creating employment but building an economy that could address histories of colonization and extractivism to support self-sufficiency, sustainability, and healthy communities.

"[G]overnments do make bad decisions, whether they intend to or not, and corporations do not have the communities' best interest at heart." (N-12)

Resistance to the PNW LNG proposal was deeply grounded in local communities' desire to protect that which they value—yet it was also grounded in the failure of the government to respond to their concerns, which were effectively stonewalled: there was no forum in which they could be meaningfully considered and potentially addressed, and all levels of government approved the project without concession.

Failures of governance extended to both industry and government-led decision-making processes. Interviewees emphasized how difficult it was for their communities to access information about the LNG industry when projects were first being proposed in the region. LNG was a brand-new industry for the region and people were not familiar with the costs and benefits associated with it, yet they struggled to find the information they sought: "*Right from the start the onus was on the community and for individual people who have lives, who are raising children, who have jobs, to educate yourself—through what source? [...] The government isn't giving us the information*" (PR-9). Instead of offering unbiased information about LNG, the government primarily relied on industry to inform the public about the project, which left people confused and suspicious: "*All of us were scratching our heads trying to really put a figure on what the real jobs were, what the real economy that was being derived from [the LNG industry], and it was an impossible task because the misinformation, the misdirection was such that it was very hard to get the information that would actually provide you that*" (PR-1).

In addition to the lack of information available, interviewees emphasized the lack of a public forum in which to learn about and discuss the proposals; in speaking about their experience attending an information session about an LNG proposal for the region, an interviewee described the session as *"yet another forum for industry to present its point of view"* (PR-7). Interviewees unsuccessfully requested that government change the industry-led model of public meetings: *"we've been asking for more engagement, for them to not let proponents do these open houses anymore ... For the BC environmental assessment office to host public meetings and have it be a public forum, a true public forum around projects"* (N-13). These frustrations were exacerbated by a perception that the provincial government had already made up its mind and had no interest in the communities' concerns: *"People felt really disempowered by LNG because government wasn't just backing it, they were pushing it. Come hell or high-water, LNG was happening. People ... didn't feel like they had a voice, they felt like it was going to happen whether people wanted it to or not"* (N-12).

These concerns became most acute when the Environmental Assessment (EA) process failed to highlight potential impacts of the project on salmon populations: "*Resources Canada [was] saying* 'no significant damage to the salmon populations if we destroy the Flora Bank.' *And all the First*

Nations went 'are you kidding?' *And the sedimentologist hired by Lax Kw'alaams said* 'are you kidding?' ... *Every scientist who was consulted knew there was something suspicious about that finding"* (S-17). Concerns expressed by an increasing number of independent scientists about the science guiding the EA—and thus formal decision-making-process[4]—enhanced community members' concerns, and solidified their resistance: *"It became really blatant in terms of the Petronas project as to how the science was being conducted and what poor standard it was being conducted. So it became a focal point for us - those who were opposing it - to go after the continued lack of science, or the continued depth of the science, or the lack of utilizing scientific principles and protocol"* (PR-1). The absence of attention to risks highlighted by both local and scientific communities deepened the communities' perceptions that the decision-making process was deeply broken, and that the decision was being motivated by political concerns rather than grounded in science and concern for community wellbeing. As one interviewee put it: *"independent science is key. There's no faith in it because it's all proponent-led and driven. And we've seen a lot of manipulation - Pacific NorthWest LNG was a classic example of incompetent science and manipulating models and that sort of stuff to get the outcome they wanted"* (N-13).

In this way, flaws in the approval processes, the exclusion of effective public engagement, and a decision-making process that failed to respond to community needs and priorities were defining grounds for opposition to the PNW LNG project. Concerns about environmental and economic impacts of the proposed project, and more specifically about the perceived distribution of risks and benefits, may well have been surmountable in this case because there was a considerable appetite for new economic initiatives in the region and some degree of openness to LNG. However, rather than responding to these concerns, the decision-making processes around the proposal ignored or rejected them, and this failure of decision-making processes was decisive in mobilizing interviewees' opposition to the project.

[4] These concerns were widely shared within the scientific community; for example, B. Jang and McCarthy (2016) write: "The CEAA draft report for the Pacific NorthWest LNG project is a symbol of what is wrong with environmental decision-making in Canada. An obvious risk of a flawed assessment is that it will arrive at an incorrect conclusion [...] CEAA did not adequately consider decades of scientific research on salmon in the Skeena River estuary and instead relied on proponent-funded studies that were substantially more limited in scope and duration and that reached different conclusions compared to the larger body of available science."

Discussion and conclusions

What arises most profoundly from this case are the failures of governance, failures that extend deep into the past—including colonization—through the more recent mismanagement of the supposedly renewable resource industries on which communities depend, and right into the present case of LNG. Despite concerns from both local and scientific communities, the project received full approval from all levels of government. The failure of decision-making processes to respond to these concerns reinforced interviewees' perception that governments neither understood nor were willing to protect their communities, a finding that is not uncommon in studies of resistance to new energy infrastructure (Shaw et al., 2015). The legacy in this case is a failed project, but more: it leaves communities who are struggling to navigate processes of economic transition feeling alienated, without adequate support from and thus trust in governments.[5]

The stage was set for this failure of governance early on, when then-Premier Christy Clark dismissed resistance to LNG as "the forces of no," painting them as NIMBY protestors who were opposed to progress of all kinds, demonizing their genuine concerns about the project's effects on their home and community, and stigmatizing their desire for a better future (Gill, 2016). What closer attention to the motivations of the resistance in this case reveals, however, is people's commitment to place and community. Devine–Wright's (Devine-Wright, 2011, 2013) work emphasizes the importance of place attachment—the emotional bonds people have to places—when examining community responses to energy projects. Stedman (2002) further asserts that "we are willing to fight for places that are central to our identities" (p. 577). These emotional bonds speak to community values, and the desire to protect those values. As one interviewee put it: *"I went to all the community hearings and it was kind of an eye opener for me too, to see how people feel about this place, the deep connection people have to the place"* (N-10). This commitment was not only or primarily economic; it

[5] Not surprisingly, subsequent to this research a populist proindustry movement has emerged in the region. Grounded in Kitimat, a community that supported and is now experiencing the construction "boom" of an LNG plant similar to the PNW LNG proposal, the movement builds on and from the frustration expressed by those who resisted PNW LNG, but toward a vision of communities supported by extractive industries. This movement, and the dynamics it expresses, will make it much more challenging to realize the nascent vision and emergent alliances that were expressed in the opposition to the PNW LNG proposal.

expressed a commitment to local histories, cultures, ecologies, and relationships, all documented as key sources of resistance more widely (Bell et al., 2013; Taylor, 2008; van der Horst & Vermeylen, 2011). The government's aggressive promotion of the LNG industry, a stance that continued when a new government took over from Clark's Liberals in 2017, exposed the lack of other programs, plans, or resources to better support these communities to restructure their economies and adapt to the realities they face, including the challenges posed by necessary economic and energy transitions.

What might more effective governance look like? Governments can and should ensure that decision-making processes are grounded in robust science and effective community engagement and are not characterized by asymmetric power relations in decision-making. Major projects such as the PNW LNG raise community-wide discussions about a region's economic future but in narrow terms. At a minimum, communities should be provided with the space and resources to frame these projects within a wider development context, including the desire expressed with regards to this project not just for a different industry, but a different kind of economy—one that would sustainably support local communities. Region-wide planning initiatives could better prepare communities to engage with economic opportunities at all scales (Hayter & Nieweler, 2018; Markey et al., 2012). Decision-making processes should mitigate potentially negative community impacts and to ensure that benefits will flow to communities (Smith, 2019). In this case, foregrounding community values and priorities might not have resulted in a different outcome for the specific project. However, it would have laid a better foundation for engaging the more fundamental challenge the region, and the province, face: that of weaning the economy off of the excessive exploitation of the ecosystems necessary to support rural flourishing, and supporting a transition to a low-carbon future.

The challenges of economic transition are profound, and especially so for rural communities whose distance from urban centers constrains options, not only in Northern British Columbia but also in many regions around the world. Against this backdrop, community-scale resistance to protect the integrity of the ecosystems on which the future of these communities hinges—at the potential cost of short term economic investment in their region—is especially poignant. That the response of the government of the day was to dismiss and override these concerns and proceed aggressively—based on insufficient and problematic decision-making processes—to approve the project, reinforced the past history in the region of state-led unsustainable extractive development, and a pattern of alienation of rural communities from the government. This raises troubling implications for the future, as the region struggles with economic transition and heads into a period where strong, government-led actions to mitigate climate change are likely necessary.

References

Allen, M. (2018). *Part 1: Death of retail in Prince Rupert*. The northern view. https://www.thenorthernview.com/business/part-1-death-of- retail-in-prince-rupert/.

Arsel, M., Hogenboom, B., & Pellegrini, L. (2016). The extractive imperative and the boom in environmental conflicts at the end of the progressive cycle in Latin America. *Extractive Industries and Society, 3*(4), 877–879. https://doi.org/10.1016/j.exis.2016.10.013

Barnes, T. J., & Hayter, R. (1997). *Troubles in the rainforest: British Columbia's forest economy in transition*. Western Geographical Press.

BC Oil and Gas Commission. (2020). *Prince Rupert gas transmission*. https://www.bcogc.ca/public-zone/major-projects-centre/prince-rupert-gas-transmission.

Bell, D., Gray, T., Haggett, C., & Swaffield, J. (2013). Re-visiting the "social gap": Public opinion and relations of power in the local politics of wind energy. *Environmental Politics, 22*(1), 115–135. https://doi.org/10.1080/09644016.2013.755793

Biersack, A. (2006). Reimagining political ecology: Culture/power/history/nature. In A. Biersack, & J. B. Greenberg (Eds.), *Reimagining political ecology* (pp. 97–120). Duke University Press.

Blomley, N. (1984). "Shut the province down": First Nations blockades in British Columbia. *BC Studies, 111*, 1984–1995.

Bowles, P., & MacPhail, F. (2017). The town that said "No" to the Enbridge Northern Gateway pipeline: The Kitimat plebiscite of 2014. *The Extractive Industries and Society, 4*(1), 15–23. https://doi.org/10.1016/j.exis.2016.11.009

Bowles, P., & Veltmeyer, H. (2014). *The answer is still no: Voices of pipeline resistance*. Fernwood Publishing.

Cashore, B. W. (2014). *In search of sustainability: British Columbia forest policy in the 1990s*. UBC Press.

Cattaneo, C., & Morgan, G. (2017). Pacific NorthWest LNG, 2012–2017: How to kill an LNG project in Canada. *Financial Post*. https://business.financialpost.com/commodities/energy/pacific-northwest-lng-2012-2017-how-to-kill-an-lng-project-in-canada.

Darwish, L. (2014). *LNG pipedreams, fractured futures and community resistance*. Council of Canadians. https://canadians.org/blog/lng-pipedreams-fractured-futures-and-community-resistance.

Devine-Wright, P. (2011). Place attachment and public acceptance of renewable energy: A tidal energy case study. *Journal of Environmental Psychology, 31*(4), 336–343. https://doi.org/10.1016/j.jenvp.2011.07.001

Devine-Wright, P. (2013). Explaining"NIMBY"objections to a power line: The role of personal, place attachment and project-related factors. *Environment and Behavior, 45*(6), 761–781. https://doi.org/10.1177/0013916512440435

Friends of Wild Salmon. (n.d.). *The Lelu island declaration*. Retrieved January 20, 2020, from: http://friendsofwildsalmon.ca/campaigns/detail/liquefied_natural_gas_lng_development/the_lelu_island_declaration/.

Garvie, K. H., Lowe, L., & Shaw, K. (2014). Shale gas development in fort Nelson First Nation territory: Potential regional impacts of the LNG boom. *BC Studies, 184*, 45–72. https://doi.org/10.14288/bcs.v0i184.184887

Garvie, K. H., & Shaw, K. (2014). Oil and gas consultation and shale gas development in British Columbia. *BC Studies, 184*, 73–102. https://doi.org/10.14288/bcs.v0i184.184888

Garvie, K. H., & Shaw, K. (2016). Shale gas development and community response: Perspectives from treaty 8 territory, British Columbia. *Local Environment, 21*(8), 1009–1028. https://doi.org/10.1080/13549839.2015.1063043

Ghoussoub, M. (2017). *Pacific north west LNG project in Port Edward, B.C., no longer proceeding*. CBC News. https://www.cbc.ca/news/canada/british-columbia/pacific-northwest-lng-project-in-port-edward-b-c-no-longer-proceeding-1.4220936.

Gilchrist, E. (2017). Fact checking Christy Clark's LNG claims. *The Narwhal*. https://thenarwhal.ca/fact-checking-christy-clark-s-lng-claims/.

Gill, I. (2016). *Scolding BC's "forces of no," our premier crassly divides us*. The tyee. https://thetyee.ca/Opinion/2016/02/06/Forces -No-Premier-Divides-Us/.

Harris, C. (2002). *Making native space: Colonialism, resistance, and reserves in British Columbia*. UBC Press.

Hayter, R., & Nieweler, S. (2018). The local planning-economic development nexus in transitioning resource-industry towns: Reflections (mainly) from British Columbia. *Journal of Rural Studies, 60*, 82–92. https://doi.org/10.1016/j.jrurstud.2018.03.006

Horne, G. (2015). *The stand at Lelu Island: B.C. First Nations vow to halt LNG project*. Ricochet Media. https://ricochet.media/en/730/the-stand-at-lelu-island-bc-first-nations-vow-to-halt-lng-project.

Horne, M., & MacNab, J. (2014). *LNG and climate change: The global context*. Pembina Institute. https://www.pembina.org/pub/lng-and-climate-change-the-global-context.

van der Horst, D., & Vermeylen, S. (2011). Local rights to landscape in the global moral economy of carbon. *Landscape Research, 36*(4), 455–470. https://doi.org/10.1080/01426397.2011.582941

Hughes, E. (2017). New evidence of contaminants from fracking. *Canadian Medical Association Journal: Canadian Medical Association Journal, 189*(31), E1025–E1026. https://doi.org/10.1503/cmaj.1095459

Hume, M. (2016). Contrary to Clark's belief, opposition to LNG project is about science. *The Globe and Mail*. https://www.theglobeandmail.com/news/british-columbia/contrary-to-clarks-belief-opposition-to-lng-project-is-about-science/article29204458/.

Jang, T. (2017). *Who owns the land — the people or the chief?* Discourse media. https://www.thediscourse.ca/reconciliation/who-owns-the-land-the-people-or-the-chief.

Jang, B., & McCarthy, S. (2016). Scientists urge Catherine McKenna to reject Pacific NorthWest LNG report. *The Globe and Mail*. https://www.theglobeandmail.com/news/british-columbia/scientists-urge-catherine-mckenna-to-reject-pacific-northwest-lng-report/article29093139/.

Keeling, A., & Sandlos, J. (2016). Introduction: Critical perspectives on extractive industries in Northern Canada. *Extractive Industries and Society, 3*(2), 265–268. https://doi.org/10.1016/j.exis.2015.10.005

Kelly, A., & Morgan, B. (2016). *Divide and conquer*. Discourse Media. https://thediscourse.ca/urban-nation/divide-and-conquer.

Leadnow. (n.d.). *Petronas LNG - BC's carbon bomb*. Retrieved January 13, 2020, from: https://www.leadnow.ca/petronas-carbon-bomb/.

Linnett, C. (2016). Federal government hit with multiple legal challenges against Pacific NorthWest LNG project. *The Narwhal*. https://thenarwhal.ca/federal-government-hit-multiple-legal-challenges-against-pacific-northwest-lng-project/.

Lough, S. (2017). *The footprint Pacific NorthWest LNG left behind*. The northern view. https://www.thenorthernview.com/news/the-footprint-pacific-northwest-lng-left-behind/.

Lutz, J. S. (2008). *Makúk: A new history of aboriginal-white relations*. UBC Press.

Magnusson, W., & Shaw, K. (2003). *A political space: Reading the global through clayoquot sound*. University of Minnesota Press.

Markey, S., Halseth, G., & Manson, D. (2012). *Investing in place: Economic renewal in Northern British Columbia*. UBC Press.

McClymont, K., & O'Hare, P. (2008). "We're not NIMBYs!" Contrasting local protest groups with idealised conceptions of sustainable communities. *Local Environment, 13*(4), 321–335. https://doi.org/10.1080/13549830701803273

Menzies, C. R. (2015). Oil, energy, and anthropological collaboration on the northwest Coast of Canada. *Journal of Anthropological Research, 71*(1), 5–21. https://doi.org/10.3998/jar.0521004.0071.101

Menzies, C. R., & Butler, C. F. (2008). The Indigenous foundation of the resource economy of BC's north coast. *Labour/Travail, 61*, 131–149.

Northwest Institute. (2019). *Proposed liquefied natural gas (LNG) projects in northern B.C.* http://northwestinstitute.ca/images/uploads/LNG_Tables_Sept10_2019.pdf.

Parfitt, B. (2017). *Fracking, First Nations and water respecting Indigenous rights and better protecting our shared resources*. Canadian Centre for Policy Alternatives. https://www.policyalternatives.ca/sites/default/files/uploads/publications/BC%20Office/2017/06/ccpa-bc_Fracking-FirstNations-Water_Jun2017.pdf.

Pembina Institute. (2016). *Pacific NorthWest LNG could become largest carbon polluter in Canada*. https://www.pembina.org/media-release/pacific-northwest-lng-could-become-largest-carbon-polluter-in-canada.

Petronas. (2017). *PETRONAS and partners will not proceed with Pacific NorthWest LNG project [press release]*. https://www.petronas.com/media/press-release/petronas-and-partners-will-not-proceed-pacific-northwest-lng-project.

Price, M. H. H., Darimont, C. T., Temple, N. F., & MacDuffee, S. M. (2008). Ghost runs: Management and status assessment of Pacific salmon (*Oncorhynchus* spp.) returning to British Columbia's central and north coasts. *Canadian Journal of Fisheries and Aquatic Sciences, 65*(12), 2712–2718. https://doi.org/10.1139/F08-174

Renshaw, T. (2014). *LNG goes global; B.C. gas export opportunities go sideways*. Business in Vancouver. https://biv.com/article/2014/12/lng-goes-global-bc-gas-export-opportunities-go-sid.

Shaw, K., Hill, S. D., Boyd, A. D., Monk, L., Reid, J., & Einsiedel, E. F. (2015). Conflicted or constructive? Exploring community responses to new energy developments in Canada. *Energy Research and Social Science, 8*, 41–51. https://doi.org/10.1016/j.erss.2015.04.003

Sierra Club of BC. (n.d.). *Say no to LNG on Lelu Island*. Retrieved January 13, 2020, from: https://sierraclub.bc.ca/say-no-lng-lelu-island/.

Smith, C. (2019). *Crystal Smith: Haisla supports gas pipeline because it means opportunities for First Nations*. Vancouver Sun. https://vancouversun.com/opinion/op-ed/crystal-smith-haisla-supports-gas-pipeline-because-it-means-opportunities-for-first-nations.

Spice, A. (2018). Fighting invasive infrastructures. *Environment and Society, 9*(1), 40–56. https://doi.org/10.3167/ares.2018.090104

Stedman, R. C. (2002). Toward a social psychology of place: Predicting behavior from place-based cognitions, attitude, and identity. *Environment and Behavior, 34*(5), 561–581. https://doi.org/10.1177/0013916502034005001

Stephenson, E., Doukas, A., & Shaw, K. (2012). Greenwashing gas: Might a "transition fuel" label legitimize carbon-intensive natural gas development? *Energy Policy, 46*, 452–459. https://doi.org/10.1016/j.enpol.2012.04.010

Stokes, D. M., Marshall, B. G., & Veiga, M. M. (2019). Indigenous participation in resource developments: Is it a choice? *Extractive Industries and Society, 6*(1), 50–57. https://doi.org/10.1016/j.exis.2018.10.015

Taylor, K. (2008). Landscape and memory: Cultural landscapes, intangible values and some thoughts on Asia. In *16th ICOMOS general assembly and international symposium: "Finding the spirit of place – between the tangible and the intangible"*. http://openarchive.icomos.org/139/.

Veltmeyer, H., & Bowles, P. (2014). Extractivist resistance: The case of the Enbridge oil pipeline project in Northern British Columbia. *Extractive Industries and Society, 1*(1), 59–68. https://doi.org/10.1016/j.exis.2014.02.002

Walters, C. J., Lichatowich, J. A., Peterman, R. M., & Reynolds, J. D. (2008). *Report of the Skeena independent science review panel.* A report to the Canadian Department of Fisheries and Oceans and the British Columbia Ministry of the Environment. https://salmonwatersheds.ca/libraryfiles/lib_b_157.pdf.

Wilderness Committee. (2016). *Ottawa approves Pacific NorthWest LNG despite major opposition.* https://www.wildernesscommittee.org/news/ottawa-approves-pacific-northwest-lng-despite-major-opposition.

Wilson, J. (1998). *Talk and log: Wilderness politics in British Columbia.* UBC Press. https://doi.org/10.2307/3985548

de Wolff, A., & Broten, D. (Eds.). (2020). *All fracked up: The costs of LNG to BC.* Watershed Sentinal Books.

Zussman, R. (2016). *B.C. Premier Christy Clark still trying to deliver on her LNG promise.* CBC News. https://www.cbc.ca/amp/1.3436887.

CHAPTER 12

Impact geographies of gas terminal development in the northern Australian context: insights from Gladstone and Darwin

Claudia F. Benham
School of Earth and Environmental Sciences, The University of Queensland, St Lucia, QLD, Australia

Introduction

In the last decade, the gas industry in Australia has significantly expanded Liquefied Natural Gas (LNG) infrastructure across regional or remote areas within the states of Queensland, the Northern Territory, and Western Australia—the country's northern periphery (Eikeland et al., 2016). This expansion has given rise to new geographies of impact created by the interaction between gas development and regional communities (Haggerty et al., 2018). Gas projects can inject significant funds into regional economies and provide lasting community infrastructure, but the industry is also associated with negative socioeconomic impacts including labor and housing shortages in the surrounding region, increased living costs, traffic, and crime (Measham, 2016). It is also increasingly apparent that large-scale gas development can involve psychosocial impacts, including disruption to residents' sense of, and attachment to, place and community (Jacquet & Stedman, 2014).

While scholars have long been attentive to the socioeconomic impacts of boomtown development (see, for example, Gilmore, 1976), interest in affect is more recent. Studies on coal mining (Albrecht et al., 2007), wind and tidal energy (Devine-Wright, 2005, 2011), shale gas development (Jacquet & Stedman, 2014) and coal seam gas mining (McCrea et al., 2019) suggest that psychosocial factors such as place attachment and impacts on well-being play a critical role in shaping local attitudes toward new energy projects. Place attachment refers to the emotional bonds people form with

Public Responses to Fossil Fuel Export
ISBN 978-0-12-824046-5
https://doi.org/10.1016/B978-0-12-824046-5.00009-6

valued places and communities, as well as the process by which these bonds are created (Devine-Wright & Quinn, 2020). Industrial developments that markedly change the landscape or induce major social and economic disruption can also rupture local place attachments and lead to solastalgia—emotional distress associated with loss of the environment or valued places (Albrecht et al., 2007). The speed of gas development, as well as community concerns about its risks to health, property and the environment, and impacts on community character, increases the likelihood of psychosocial disruption (Jacquet, 2014).

This chapter adopts the concept of "impact geographies" (Haggerty et al., 2018) as a means of understanding how place-based factors influence local community perceptions of the gas industry, and relatedly, its impacts on local quality of life and subjective well-being. The term refers to "a spatially-bounded area that features a distinct constellation of historical, physiographic (including climate, geology and ecology), economic, and cultural factors that influence the nature of oil and gas development and the character and magnitude of its impacts on local people, ecologies and landscapes" (Haggerty et al., 2018, p. 621). Although originating in disparate research fields, the impact geographies framework reflects the principles of social–ecological systems frameworks (Armitage et al., 2012) that define human well-being as a state of "being with others and the natural environment" in a way that meets human needs and allows people to experience satisfaction with their way of life (Armitage et al., 2012, p. 3). Viewing gas development as embedded within an established social–ecological system "enables expanding the scope of existing research on the impacts of change" (Lai et al., 2017) to reflect the full scope of the industry's effects at the local scale.

As Lai et al. (2017, 492) observe, well-being in gas-adjacent communities is "highly dependent on rural residents' interactions with the social and ecological systems within the communities, and the various personal and shared community resources that support the needs and goals of individuals and groups, and their preferred way of life." Many coastal residents in northern Australia hold strong attachments to place through Aboriginal and Torres Strait Islander peoples' relationships to country (Garnett et al., 2009), resource-dependent livelihoods and lifestyles, and a high degree of interaction or identification with iconic environments such as the Great Barrier Reef (GBR) (Marshall et al., 2013). Regional communities in northern Australia are also experiencing social and economic changes associated with ageing populations, pressure to sustain regional economies, and a high level of exposure to the impacts of climate change and biodiversity loss (Chambers et al., 2018; Dale, 2013). Taken as a whole, these factors create a unique context within which community interactions with the gas industry must be understood (see Luke et al., 2018).

LNG development in Australia

Approximately 1.87% of global proven conventional gas reserves and 4% of global unconventional gas reserves are located in Australia (Geoscience Australia, 2015). Although Australia has exported LNG since 1989, rapid expansion in LNG terminal development has occurred since 2011, driven by technological developments and market conditions that enabled the accessibility of previously undeveloped Coal Seam Gas (CSG) resources, and the reorientation of the industry toward the export market (Liu et al., 2020). The acceleration of gas terminal development in Australia since 2011 reflects a global trend toward increasing proliferation of LNG developments, even as market conditions have softened during the COVID-19 pandemic (Bresciani et al., 2020). In the year to May 2020, "the fossil fuel industry worldwide ... more than doubled the amount of liquefied natural gas (LNG) terminal capacity under construction" (Plante et al., 2020).

Ten LNG terminals are currently operational in Australia, and these are located in the three northern-most states of Queensland, Western Australia, and the Northern Territory. In 2016, Australian Government environmental reporting noted the substantial increase in mining activities in these three states, and the historically high number of port developments built to accommodate the export of coal, gas, and minerals (Clark & Johnston, 2017). Australian LNG exports have increased, with some fluctuations, since 2003, consolidating the country's position as one of the two largest global exporters of LNG (Liu, Shi, & Laurenceson, 2020; Towler et al., 2016; Geoscience Australia, 2021).

The majority of Australian research on local perceptions of gas development has focused on gas fields, largely in the eastern states of Queensland and New South Wales (Luke et al., 2018). This research indicates that gas development is associated with boom-and-bust cycles in local housing markets, social disruption, and income inequality (Brueckner et al., 2013; Fleming & Measham, 2015; Rolfe et al., 2007). It also suggests that community attitudes toward gas development can differ between regional Australian contexts (McCrea et al., 2020) and within individual communities

(Lai et al., 2017). As in other global contexts, local attitudes toward gas field development in Australia are shaped largely by the social, environmental, and economic risks of a project, trust in project governance, and fair distribution of project benefits (McCrea et al., 2020; Moffat et al., 2014, 2018). Residents' acceptance of gas development largely depends on their ability to capitalize on the livelihood benefits of a gas boom while mitigating negative impacts on well-being (Haggerty et al., 2018).

Purpose of this chapter

This chapter explores the relationship between place, community, and gas development through case studies of gas terminal construction in Gladstone and Darwin, key hubs for the Australian LNG industry. The chapter identifies place-based factors shaping community attitudes toward LNG terminal development in Gladstone during LNG plant construction (2011 to 2014), and discusses their implications for the psychosocial wellbing of local residents. The chapter also evaluates the case of Gladstone against the impacts of LNG development in Darwin, a larger northern Australian city where a gas terminal was constructed between 2014 and 2018, using a comparative case study approach to draw out similarities and differences (see Krehl and Weck, 2020).

Methods

The chapter draws on case study research conducted by the author in 2014 (Gladstone) and 2018 (Darwin). In Gladstone, this work involved a mail-out survey of 296 local residents randomly selected from local Council records, two focus groups conducted with survey respondents, and in-depth interviews with 43 key stakeholder representatives, including Traditional Owners (see Benham & Daniell, 2016 for a comprehensive discussion of the methods used in the Gladstone case study). In Darwin, the case study involved a mail-out survey of 246 local residents, randomly selected through the Australia Post address database. Survey data were supplemented by a review of peer-reviewed literature, industry, and government reports, and media articles related to the two case studies, published between 2011 and 2020. Due to the lack of primary interview data from Darwin, the latter case study draws strongly on secondary sources. While not an exhaustive analysis, the aim of this chapter is to illuminate the impact of social, cultural, and environmental factors on community attitudes toward gas development in two different regional contexts, and to discuss the implications for future, similar, developments.

Gas terminal development in Australia: a tale of two cities

Gladstone, Queensland

Located along the coast of the GBR, the region and port of Gladstone is a key center for industrial processing. Approximately 62,000 people live in the Gladstone region (Australian Bureau of Statistics, 2017a), which includes the city of Gladstone, where the port is located, and several smaller towns and rural smallholdings. Population estimates suggest that more than 5000 people in the region are employed full-time in industries such as alumina refinement, aluminum smelting, coal export, cement and chemicals production, and power generation (Queensland Government Statistician's Office, 2016).

Heavy industry has played a key role in the regional economy of Gladstone since the construction of an aluminum smelter in 1967, when "the influx of people required for the construction of the aluminium smelter transformed a small quiet town that had unsealed roads, no sewerage and limited community facilities" (Barker, 1981 in Smith and Kelly, undated, 3). The region has experienced economic peaks and troughs associated with the entrance of new industries over several decades, but the largest "boom" occurred between 2011 and 2014, due to the simultaneous construction of three LNG terminals and a coal export terminal (Benham 2016; Smith and Kelly, undated). LNG infrastructure in Gladstone comprises three separate gas liquefaction facilities—the Gladstone Liquefied Natural Gas project (GLNG); the Australia Pacific Liquefied Natural Gas company (APLNG); and the Queensland Curtis Liquefied Natural Gas (QCLNG). All three projects were constructed to process CSG from the Surat and Bowen Basins in inland Queensland. LNG infrastructure in the Port of Gladstone comprises three, co-located gas liquefaction facilities; gas and water pipelines, and shipping facilities. The three terminals, comprising six LNG trains in total, are located close to shore on Curtis Island, within the Port of Gladstone. Capital dredging was undertaken to remove 36 million cubic meters of port sediments, in order to create shipping channels and swing basins for deeper-draft LNG tankers (GHD, 2009).

The population of Gladstone boomed between 2011 and 2014 as the construction of the three LNG plants coincided with the final stages of construction on a nearby coal terminal expansion (the Wiggins Island Coal Export Terminal (WICET) see Benham, 2016). A workforce of approximately 14,500 was employed during peak construction of the three Gladstone LNG facilities, including a large Fly-In Fly-Out workforce (Bechtel, undated). The LNG construction phase in Gladstone led to rapid social and economic changes in the surrounding region:

> *The Gladstone community moved rapidly from a close knit family orientated environment into a transient population, with an influx of over 15,000 workers arriving. Although Development Permits were in place for subdivisions and residential development, the construction of these developments became problematic due to the lack of accommodation for the indirect workers as well as obtaining and retaining a work force in existing businesses. The population boom drained the housing rental market with a vacancy rate of 0.8% and the cost of living soared. People were living in caravans and in their cars, with motels and hotels 'hot bedding' the shift workers. People were also sharing Dwelling Houses with up to 12 people reported to be living in a single residence*
>
> ***Smith and Kelly (undated, 2).***

Other impacts included the displacement of low- and middle-income earners, such as retired residents, teachers, nurses, and police, who were unable to secure affordable housing (Mitchell-Whittington, 2017). Residents also reported feeling a loss of sense of community as large numbers of LNG workers moved into the town and other residents were displaced (Benham, 2016). Local authorities sought to relieve pressure on the local housing market by releasing land for housing development as the industry completed a 6000-bed worker accommodation complex approved by the Queensland and Australian governments, leading to a rapid decline in house prices (Benham, 2016; Smith and Kelly, undated). Drug, alcohol, and domestic violence services reported an increase in demand as workers and families faced multiple pressures including long hours, a male-dominated workforce culture and periods of separation due to fly-in fly-out work schedules (Benham, 2016). The social and economic impacts of LNG construction led media to report that "the biggest loser" of the Australian gas boom was Gladstone itself (Mitchell-Whittington, 2017).

Community concern about the impacts of the industry on the GBR also played a role in influencing local perceptions of LNG in Gladstone. The Reef is regarded as Australia's "most inspiring" icon (Goldberg et al., 2016), endowed with rich biodiversity and cultural importance. In Gladstone, concerns about the risks of LNG development to local marine ecosystems were a key driver of local attitudes toward the industry, even among those who did not normally hold strong pro-environmental values (Benham, 2016). Community concerns focused largely on the impact of dredging on commercial and recreational fisheries species, marine turtles, and dugong within the Port of Gladstone, and the impact of spoil disposal on coral reef ecosystems within the broader Great Barrier Reef World Heritage Area (see Llewellyn et al., 2019). Some residents reported experiencing distress about loss of access or amenity associated with meaningful sites or activities, such as recreational fishing or sailing sites in the port, or camping sites in Curtis Island National Park and Conservation Park (Benham, 2016). Concern was highest among residents who held strong personal attachments to the local marine environment or who perceived that LNG development had been the key factor causing the local ecosystem to change from a more "pristine" environment to a polluted, industrialized area, a perception that was described by some as grief for the lost environment (Benham, 2016). This view reflects broader trends in affective vulnerability to environmental loss in the GBR (Marshall et al., 2019).

In general, however, there was relatively little local community organizing against LNG development in Gladstone, in comparison to significant mobilization of gas field communities in Queensland and elsewhere in Australia (see Luke et al., 2018). This may be due in part to community familiarity or involvement with heavy industry. Previous research has indicated that perceiving new energy development as a continuation of the past can lead to greater community acceptance (Devine-Wright, 2009), and more recently, that communities with an industrial history demonstrate greater acceptance of new gas developments (McCrea et al., 2020). A lack of community organizing can also reflect a perception that little can be done to change the course of development (Devine-Wright & Quinn, 2020) – a perception expressed by some stakeholders in Gladstone (Benham & Hussey, 2018). Australia's governance of the Gladstone LNG projects attracted international criticism due to its failure to keep up with the rapid pace and scale at which the projects were being built. The potential impacts of coastal development on the GBR led the United Nations Educational, Scientific and Cultural Organization (UNESCO) to state that the "scale and pace of [port] development proposals [in the Great Barrier Reef region] appear beyond the capacity for independent, quality and transparent decision making" (Douvere & Badman, 2012).

Although residents in the Gladstone Region were largely familiar with industrial development prior to the construction of the three LNG terminals, the potential impacts of LNG development on the GBR, and the speed and scale of gas development in the region, drove community concern and led to the reporting of psychosocial impacts—notably disruptions to local identities, quality of life, and attachment to important places (Benham, 2016). Outdoor lifestyles are highly valued by residents living along the coast of the GBR, and the reef holds extremely high biodiversity, cultural and recreational values that, when threatened, are associated with emotional distress and grief responses (Marshall et al., 2019). The perception that fisheries, recreational areas or critical local ecosystems such as turtle nesting sites on Curtis Island were threatened by environmental changes associated with gas development in the Port of Gladstone led to feelings of distress and grief among some local residents (Benham, 2016). These impacts are typical of psychosocial disruptions associated with energy developments that induce concern or distress about changes to valued places (Albrecht et al., 2007). The size of the LNG workforce relative to the Gladstone community also induced significant disruption to local social networks as previous residents moved out of the region and large numbers of new residents or temporary workers moved in. Adding to this, the speed at which the industry developed from regulatory approvals to construction provided relatively little time for surrounding communities and local authorities to adapt (Benham, 2016).

Darwin, Northern Territory

The experience of the community in Gladstone can be compared with another LNG terminal constructed in Australia's north between 2014 and 2018. The Icthys LNG facility is located in the Port of Darwin, Australia's northern-most capital city. The facility was built by oil and gas company Inpex to process and export gas from the Icthys gas field, a conventional gas resource located in the Browse Basin off the coast of Western Australia (WA), and includes two LNG trains. Construction on the facility began in 2013, and included a major dredging campaign to remove approximately 16 million cubic meter of sediment and hard rock to allow shipping movements (Carle et al., 2015). The Inpex plant is the second LNG facility in the Port of Darwin, with an earlier facility, the Darwin LNG project, operational since 2006. The construction of Inpex created "boomtown" effect in Darwin. An estimated 10,000 people were working on site at the height of construction (Garrick, 2018).

Darwin is home to approximately 132,000 people, including a large population of Aboriginal and Torres Strait Islander people (Australian Bureau of Statistics, 2017b). Although it is a capital city, Ennis and Finlayson (Ennis & Finlayson, 2015, p. 52) note that "a 'frontier town' identity prevails" due to Darwin's "geographic isolation, abundance of potentially deadly wildlife such as crocodiles," a large military presence, "a male-dominated, drinking culture with alcohol consumption being among the highest in the world" and a high incidence of violent crime. As in Gladstone, the large construction workforce employed for the Inpex development led to a boom town effect, which was associated with rapidly increasing housing costs, social disruption, and increasing inequality (Taylor & Carson, 2014; Lea, 2014 in Ennis et al., 2014). Inpex construction occurred as the US Marines increased their military presence in Darwin, further skewing the gender balance in the local region (Ennis & Finlayson, 2015). Social changes during the period of LNG construction were notably gendered, and included an increase in alcohol-related violence, street harassment, and increased demand for family violence support services (Ennis et al., 2014). As in Gladstone, high housing costs made it more difficult for families in crisis to find affordable accommodation (Ennis et al., 2014). Media reporting observed that "High rents, high-vis clothing and high-rise buildings are now synonymous with living in Darwin" (Curtain, 2014). Lea (2014, 224) writes that:

> *Darwin [in 2014] represents the bonanza and cataclysm that is the dazzling Inpex/military build-up. Whole buildings in the many new high-rise complexes in and around the old peninsula are filled with strangers, non-resident workers the fastest-growing demographic ... Darwin offers work, but not the hope of a home, the toll price for living in a boomtown. It has become a prosperous, unequal city*
>
> ***Lea (2014, p. 224), Ennis et al. (2016).***

Although socioeconomic disruptions associated with Inpex have been documented, local environmental concerns did not feature strongly in scholarly or media reporting. A report by Inpex suggests that the major socioecological impacts were related to the effects of dredging on local fishers (Williams & Townsend, 2015), but in a review of literature for this chapter, no other reporting of social—ecological impacts was found. This suggests that a charismatic local ecosystem such as the GBR provides a key formative point around which community concerns are shaped. A lack of documented concern for environmental impacts perhaps also reflects the transient nature of Darwin's population — which may mean fewer residents hold strong place attachments in the region — and a lack of attention to First Nations perspectives in the literature on gas development (Haggerty et al., 2018). Luke et al. (2018, 562) write that despite large Aboriginal and Torres Strait Islander populations living on country in the Northern Territory, decisions about future gas development have been made "without regard to the unique circumstances posed; mainly, that the people who will be most impacted by [gas development], are First Peoples with distinct social, political, governance, economic and spiritual lives that are inextricably linked to their ancestral lands."

Concluding reflections

Understanding community perceptions of gas developments within their social and environmental contexts, including the psychosocial impacts of development and its relationship to community well-being, is vital to minimizing the impacts of the industry and shaping future energy development in a way that is compatible with local social, environmental, and economic values, aspirations, and goals. More broadly, understanding the "impact geographies" of LNG in regional and remote contexts allows for a fuller scholarly understanding of the relationships between the environment, community, and industry in the remote "northern peripheries" that often play host to large mining and energy developments (Eikeland et al., 2016).

Although a single chapter cannot detail the full complexity of interactions between gas development and place in Australia's north, it is my hope that these case studies provide valuable insight into the geographical, environmental, sociocultural, and economic factors that shape local residents' attitudes toward gas terminal development. In Gladstone, the speed and magnitude of gas development relative to previous booms, social disruption caused by a large transient workforce, risks to highly valued GBR ecosystems, and governance shortcomings shaped local resident concerns about the development. In Darwin, where LNG development occurred on a smaller scale relative to the size of the local population, gendered impacts came to the fore, and reporting of social-environmental impacts focused largely users of Darwin Harbour, rather than on broader perceptions of place or attachments to iconic ecosystems such as the GBR. Although the socioeconomic impacts of LNG development in Gladstone and Darwin broadly reflected typical effects of 'boom town' energy development (Hall et al., 2013), local social, geographic, and cultural conditions influenced the specific ways in which these impacts were experienced 'on the ground'. Local experiences of gas development in Queensland and the Northern Territory highlight the importance of considering contextual factors and community perspectives in the planning and governance of LNG development, notably those of Traditional Owners, whose perspectives are not well-represented in scholarly reporting on gas development.

The concept of impact geographies provides an explicitly place-based framework for interpreting the attitudes of local residents toward gas development. The framework highlights the importance of contextual factors in shaping the psychosocial impacts of gas development at the local scale. Moreover, it suggests that managing future development in a way that is sensitive to local conditions, needs, and concerns beyond typical socio-economic analyses is critical if LNG development is to minimize impacts and deliver enduring benefits at the local scale.

References

Albrecht, G., Sartore, G.-M., Connor, L., Higginbotham, N., Freeman, S., Kelly, B., Stain, H., Tonna, A., & Pollard, G. (2007). Solastalgia: The distress caused by environmental change. *Australasian Psychiatry, 15*(Suppl. 1), S95–S98. https://doi.org/10.1080/10398560701701288

Armitage, D., Béné, C., Charles, A. T., Johnson, D., & Allison, E. H. (2012). The interplay of well-being and resilience in applying a social-ecological perspective. *Ecology and Society, 17*(4).

Australian Bureau of Statistics. (2017a). Australian Bureau of Statistics. https://quickstats.censusdata.abs.gov.au/census_services/getproduct/census/2016/quickstat/30805?opendocument, 2017. (Accessed 21 June 2021).

Australian Bureau of Statistics. (2017b). *2016 QuickStats, Greater Darwin.* Australian Government, Canberra. https://quickstats.censusdata.abs.gov.au/census_services/getproduct/census/2016/quickstat/7GDAR. (Accessed 21 June 2021).

Benham, C. (2016). Change, opportunity and grief: Understanding the complex social-ecological impacts of liquefied natural gas development in the Australian coastal zone. *Energy Research and Social Science, 14*, 61–70.

Benham, C. F., & Daniell, K. A. (2016). Putting transdisciplinary research into practice: A participatory approach to understanding change in coastal social-ecological systems. *Ocean and Coastal Management, 128*, 29–39. https://doi.org/10.1016/j.ocecoaman.2016.04.005

Benham, C. F., & Hussey, K. E. (2018). Mainstreaming deliberative principles in environmental impact assessment: Current practice and future prospects in the Great Barrier Reef, Australia. *Environmental Science and Policy, 89*, 176–183. https://doi.org/10.1016/j.envsci.2018.07.018

Bresciani, G., Heiligtag, S., Lambert, P., & Rogers, M. (2020). *The future of liquefied natural gas: Opportunities for growth.* https://www.mckinsey.com/industries/oil-and-gas/our-insights/the-future-of-liquefied-natural-gas-opportunities-for-growth.

Brueckner, M., Durey, A., Mayes, R., & Pforr, C. (2013). The mining boom and Western Australia's changing landscape: Towards sustainability or business as usual? *Rural Society, 22*(2), 111–124.

Carle, J., Oliver, G., Carter, S., & Usui, H. (2015). Ichthys LNG project: Successful dredging campaign in Darwin Harbour, Northern Territory. *Australasian Coasts and Ports Conference*, 147–153. https://doi.org/10.3316/informit.703324263527748

Chambers, I., Russell-Smith, J., Costanza, R., Cribb, J., Kerins, S., George, M., James, G., Pedersen, H., Lane, P., & Christopherson, P. (2018). Australia's north, Australia's future: A vision and strategies for sustainable economic, ecological and social prosperity in northern Australia. *Asia and the Pacific Policy Studies, 5*(3), 615–640.

Clark, G., & Johnston, E. (2017). *Australia state of the environment 2016: coasts, independent report to the Australian Government Minister for Environment and Energy*. Canberra: Australian Government Department of the Environment and Energy.
Curtain, Carl (2014). Darwin growth spurt fuelled by LNG. *ABC News*. https://www.abc.net.au/news/2014-04-28/darwin-lng-boom/5362818.
Dale, A. (2013). *Governance challenges for Northern Australia*.
Devine-Wright, P., & Quinn, T. (2020). Dynamics of place attachment in a climate changed world. *Place Attachment*, 226–242.
Devine-Wright, P. (2009). Rethinking NIMBYism: The role of place attachment and place identity in explaining place-protective action. *Journal of Community and Applied Social Psychology, 19*(6), 426–441.
Devine-Wright, P. (2005). Beyond NIMBYism: towards an integrated framework for understanding public perceptions of wind energy. *Wind Energy: An International Journal for Progress and Applications in Wind Power Conversion Technology, 8*(2), 125–139.
Devine-Wright, P. (2011). Place attachment and public acceptance of renewable energy: A tidal energy case study. *Journal of Environmental Psychology, 31*(4), 336–343.
Douvere, F., & Badman, T. (2012). *Mission report: Reactive monitoring mission to Great Barrier Reef (Australia) 6 to 14 March 2012*. UNESCO World Heritage Centre-IUCN.
Eikeland, S., Nilsen, T., & Taylor, A. (2016). Re-evolution of growth pole settlements in northern peripheries? Reflecting the emergence of an LNG hub in northern Australia with experiences from Northern Norway. In *Settlements at the edge*. Edward Elgar Publishing.
Ennis, G., & Finlayson, M. (2015). Alcohol, violence, and a fast growing male population: Exploring a risky-mix in "boomtown" Darwin. *Social Work in Public Health, 30*(1), 51–63. https://doi.org/10.1080/19371918.2014.938392
Ennis, G., Tofa, M., & Finlayson, M. (2014). Open for business but at what cost? Housing issues in 'boomtown' Darwin. *Australian Geographer, 45*(4), 447–464.
Fleming, D. A., & Measham, T. G. (2015). Income inequality across Australian regions during the mining boom: 2001–11. *Australian Geographer, 46*(2), 203–216.
Garnett, S. T., Sithole, B., Whitehead, P. J., Burgess, C. P., Johnston, F. H., & Lea, T. (2009). Healthy country, healthy people: Policy implications of links between indigenous human health and environmental condition in tropical Australia. *Australian Journal of Public Administration, 68*(1), 53–66. https://doi.org/10.1111/j.1467-8500.2008.00609.x
Garrick, M. (2018). *Foreign investment could fuel "massive oil and gas base" in NT within a decade, says senator*. ABC News. https://www.abc.net.au/news/2018-11-16/inpex-ichthys-lng-gas-project-foreign-investment-japan-abe/10505696.
Geoscience Australia. (2015). *Australian gas resource assessment 2015*. Australian Government, Canberra.
Geoscience Australia. (2021). *Australia's Energy Commodity Resources 2021*. Australian Government, Canberra. https://www.ga.gov.au/digital-publication/aecr2021.
GHD. (2009). *Report for Western Basin Dredging and Disposal Project, Water Quality Report, October 2009*. Gladstone Ports Corporation. https://eisdocs.dsdip.qld.gov.au/Port%20of%20Gladstone%20Western%20Basin%20Dredging/EIS/appendix-k-water-quality-report.pdf.
Gilmore, J. S. (1976). Boom towns may hinder energy resource development. *Science, 191*(4227), 535–540.
Goldberg, J., Marshall, N., Birtles, A., Case, P., Bohensky, E., Curnock, M., Gooch, M., Parry-Husbands, H., Pert, P., Tobin, R., Villani, C., & Visperas, B. (2016). Climate change, the Great Barrier Reef and the response of Australians. *Palgrave Communications, 2*(1), 15046. https://doi.org/10.1057/palcomms.2015.46

Haggerty, J. H., Kroepsch, A. C., Walsh, K. B., Smith, K. K., & Bowen, D. W. (2018). Geographies of impact and the impacts of geography: Unconventional oil and gas in the American west. *The Extractive Industries and Society, 5*(4), 619–633. https://doi.org/10.1016/j.exis.2018.07.002

Hall, N., Ashworth, P., & Devine-Wright, P. (2013). Societal acceptance of wind farms: Analysis of four common themes across Australian case studies. *Energy Policy, 58*, 200–208. https://doi.org/10.1016/j.enpol.2013.03.009

Jacquet, J. (2014). Review of risks to communities from shale energy development. *Environmental Science & Technology, 48*(15), 8321–8333.

Jacquet, J., & Stedman, R. (2014). The risk of social-psychological disruption as an impact of energy development and environmental change. *Journal of Environmental Planning and Management, 57*(9), 1285–1304.

Krehl, A., & Weck, S. (2020). Doing comparative case study research in urban and regional studies: What can be learnt from practice? *European Planning Studies, 28*(9), 1858–1876. https://doi.org/10.1080/09654313.2019.1699909

Lai, P.-H., Lyons, K. D., Gudergan, S. P., & Grimstad, S. (2017). Understanding the psychological impact of unconventional gas developments in affected communities. *Energy Policy, 101*, 492–501. https://doi.org/10.1016/j.enpol.2016.11.001

Liu, Y., Shi, X., & Laurenceson, J. (2020). Dynamics of Australia's LNG export performance: A modified constant market shares analysis. *Energy Economics, 89*, 104808. https://doi.org/10.1016/j.eneco.2020.104808

Llewellyn, L., Brinkman, R., McIntosh, E., Marshall, N., Pinto, U., Rolfe, J., & Schaffelke, B. (2019). Gladstone harbour: A case study of building social licence-to-operate in a multi-use area. *The APPEA Journal, 59*(2), 624–631. https://doi.org/10.1071/AJ18052

Luke, H., Brueckner, M., & Emmanouil, N. (2018). Unconventional gas development in Australia: A critical review of its social license. *The Extractive Industries and Society, 5*(4), 648–662. https://doi.org/10.1016/j.exis.2018.10.006

Marshall, N., Adger, W. N., Benham, C., Brown, K., Curnock, M. I., Gurney, G. G., Marshall, P., Pert, P. L., & Thiault, L. (2019). Reef grief: Investigating the relationship between place meanings and place change on the Great Barrier Reef, Australia. *Sustainability Science, 14*(3), 579–587.

Marshall, N. A., Tobin, R. C., Marshall, P. A., Gooch, M., & Hobday, A. J. (2013). Social vulnerability of marine resource users to extreme weather events. *Ecosystems, 16*(5), 797–809.

McCrea, D. R., Walton, D. A., & Jeanneret, M. T. (2020). An opportunity to say no: Comparing local community attitudes toward onshore unconventional gas development in pre-approval and operational phases. *Resources Policy, 69*. https://doi.org/10.1016/j.resourpol.2020.101824

McCrea, R., Walton, A., & Leonard, R. (2019). Rural communities and unconventional gas development: What's important for maintaining subjective community wellbeing and resilience over time? *Journal of Rural Studies, 68*, 87–99. https://doi.org/10.1016/j.jrurstud.2019.01.012

Moffat, K., Zhang, A., & Boughen, N. (2014). *Australian attitudes toward mining. Citizen Survey—2014 Results*. Australia: Commonwealth Scientific and Industrial Research Organisation, CSIRO.

Measham, T., Fleming, D., & Schandl, H. (2016). A conceptual model of the socioeconomic impacts of unconventional fossil fuel extraction. *Global Environmental Change, 36*, 101–110.

Mitchell-Whittington, A. (2017). The biggest loser of Australia's gas shortage is the city that's sucking it all up. *ABC News*. https://www.brisbanetimes.com.au/national/queensland/the-biggest-loser-of-australias-gas-shortage-is-the-city-thats-sucking-it-all-up-20170316-guzwbf.html.
Moffat, K., Lacey, J., Boughen, N., Carr-Cornish, S., & Rodriguez, M. (2018). Understanding the social acceptance of mining. In *Mining and sustainable development: Current issues* (pp. 45–62). Routledge London.
Plante, L., Browning, J., Aitken, G., Inman, M., & Nace, T. (2020). *Gas bubble: Tracking global LNG infrastructure*. https://globalenergymonitor.org/wp-content/uploads/2020/07/GasBubble_2020_r3.pdf.
W. and N. A. *Prime minister, minister for energy and emissions reduction, minister for resources* (2020). Media Release: Gas-Fired Recovery.
Queensland Government Statistician's Office. (2016). *Gladstone region population report, 2016*. Queensland Government, Brisbane. https://www.qgso.qld.gov.au/issues/3306/gladstone-region-population-report-2016.pdf.
Rolfe, J., Miles, B., Lockie, S., & Ivanova, G. (2007). Lessons from the social and economic impacts of the mining boom in the Bowen Basin 2004-2006. *Australasian Journal of Regional Studies, 13*(2), 134.
Smith, Trudy, & Kelly, Matt (undated). From tents to no tenants - Managing development when the Gladstone housing market radically changes from shortage to saturation. Gladstone Regional Council Development Services Department Planning Services Section.
Taylor, A. J., & Carson, D. B. (2014). It's raining men in Darwin: Gendered effects from the construction of major oil and gas projects. *Journal of Rural and Community Development, 9*(1), 24–40. http://umu.diva-portal.org/smash/get/diva2:967909/FULLTEXT01.pdf.
Towler, B., Firouzi, M., Underschultz, J., Rifkin, W., Garnett, A., Schultz, H., Esterle, J., Tyson, S., & Witt, K. (2016). An overview of the coal seam gas developments in Queensland. *Journal of Natural Gas Science and Engineering, 31*, 249–271. https://doi.org/10.1016/j.jngse.2016.02.040
Williams, B., & Townsend, B. (2015). Rock, sediment and fishermen—insights on community engagement from the Ichthys LNG Project dredging program. *The APPEA Journal, 55*(2), 435–435.

CHAPTER 13

Community risk or resilience? Perceptions and responses to oil train traffic in four US rail communities

Anne N. Junod
Urban Institute, Washington, DC, United States

Introduction

Due to convoluted siting and regulatory regimes governing oil and gas pipeline development in the United States, as well as growing opposition to new and existing pipeline projects (McAdam et al., 2010; Pierce et al., 2018) and pipeline capacity limitations in many regions, unprecedented volumes of crude oil have been pushed onto railroads.

Public perceptions of oil by rail (OBR) activity and exports represent a significant gap in energy development social science literatures, yet community perceptions of energy infrastructure are critical to state and federal government agencies designing and implementing regulatory and energy policy (Rabe & Borick, 2013); to long-term planning, siting, and zoning (Boudet et al., 2011; Pierce et al., 2018); and to community and other advocates in communicating potential public, economic, and environmental benefits or harms (Heikkila et al., 2014).

With the expansion of natural gas export and liquefaction terminals in coastal cities across the United States and declining domestic demand for coal coupled with increasing demand internationally, hydrocarbon exports from the United States are poised to continue through at least 2050 (United States Energy Information Administration, 2020). At least through the near and mid-term, industry will continue to rely on rail technology as a more flexible alternative to pipeline to transport resources both domestically and abroad—yet very little is known about perceptions toward OBR in communities most impacted by it.

In the years following the early 2000s Bakken oil boom, OBR shipments increased over 4000% (American Association of Railroads, 2017),

Public Responses to Fossil Fuel Export
ISBN 978-0-12-824046-5
https://doi.org/10.1016/B978-0-12-824046-5.00011-4

with much of the resource destined for coastal refineries and export terminals across the United States, Canada, and Mexico. In shale plays like the Bakken in particular—located in a remote, rural region of the United States minimally serviced by petroleum liquid and gas transmission pipelines—industry increasingly relies on the country's aging rail infrastructure to transport crude oil, which is more prone to failure or accident than pipeline (Dante et al., 2016), and related health and environmental impacts are twice the cost of pipeline (Clay et al., 2017).

Growth in OBR has increased the length of trains, with 100-car-plus "unit" trains—or trains carrying solely one commodity, often a mile or longer in length—now common (Lazo & McClain, 1996; United States Department of Transportation Pipeline and Hazardous Materials Safety Administration, 2014). In addition to being longer, these trains are heavier, more frequent, and carry more volatile hydrocarbons than were common before the shale energy boom (Lazo & McClain, 1996; United States Department of Transportation Pipeline and Hazardous Materials Safety Administration, 2014), presenting new safety and policy challenges for state and federal lawmakers.

Derailments involving oil trains have also increased, and a number of high-profile accidents—including the 2013 derailment in rural Lac-Mégantic, Québec, which killed 47 people and destroyed much of the downtown—have elevated public awareness around oil train exports in many rail communities across the United States and Canada. The Lac-Mégantic disaster, the largest of its kind in terms of human loss, has been the focus of scholarly and policy inquiry, with researchers examining subsequent social (Brisson & Bouchard-Bastien, 2018; Emmanuelle & Geneviève, 2018), public health (Mélissa et al., 2014), environmental (Santiago-Martín et al., 2015), and policy (Mark, 2018) outcomes. Other significant oil train derailments have resulted in billions of dollars of infrastructure and agricultural losses, watershed and ecosystem contamination, and related social disruptions which have been the subject of significant policy debate and news media coverage (Junod, 2020).

Even so, most scholarly analyses involve case studies specific to the Lac-Mégantic disaster, and media reporting on OBR impacts elsewhere lacks comparative and empirical grounding and tends to focus on metropolitan contexts. Overall, a significant knowledge gap exists concerning public perceptions of OBR activity and exports—a gap which stands in contrast to the growing social science examining other hydrocarbon export infrastructure and activity, including pipeline siting, natural gas development

Rail Community	Population	Rural/Urban	Hydrocarbon Derailment?	Impact or Damage
Mosier, Oregon	451	Rural	Y - 2016 Bakken Crude 16/96 cars derailed	1/3 community evacuated; 47,000 gallons spilled, 16,000 gallons vaporized or burned; 2,960 tons irremediable soil excavated; $9m+ damage
Asvhille, Ohio	4,194	Rural	N	N/A
Spokane, Washington	215,973	Urban	N	N/A
Columbus, Ohio	879,170	Urban	Y - 2012 Ethanol 17/98 cars derailed	2 injuries; 100 residences evacuated; $1.2m infrastructure damange

Figure 13.1 Community summary and experiences with oil by rail activity.

and exports (Hazboun, 2019; Pierce et al., 2018; Trang et al., 2019), and coal exports (Hazboun, 2019; McGrath, 2018).

In this chapter, I begin to address this gap in contributing novel baseline and comparative public perception findings from four rail communities across the United States with different experiences with OBR: two are in the Pacific Northwest, two are in the Midwest; two have experienced significant derailments, two have not; two are rural, two are metropolitan; all are located on major rail lines and experience comparable rates of OBR export activity (Fig. 13.1).

In addition to other "linear" energy infrastructure such as roads, transmission lines, and to a lesser extent, pipelines, little is known about community views toward OBR. This study is the first to examine community perceptions in OBR export corridors (a) cross-regionally and in both rural and urban communities; and (b) prospectively—in examining rail communities that both have and have not experienced a derailment event. This research contributes to and extends the energy export, risk perception, and environmental sociology literatures in identifying characteristics of public views about OBR, an expanding energy technology that is integral to North American energy export regimes, but has heretofore gone unexamined.

Perceptions of energy transportation and exports via rail

Limited social science focuses on the intersection of public attitudes and the transportation of energy resources. Much of the extant literature has considered perceptions of nuclear fuel and waste transportation, primarily in relation to the Yucca Mountain nuclear repository siting disputes of the late 20th century (Krannich & Albrecht, 1995; Krannich et al., 1993, pp. 263–287; National Research Council, 1984). In response to nuclear waste via rail in particular, perceptions of diminished capacity or agency in rural communities have been found to associate with increased opposition and heightened risk perceptions (Krannich & Albrecht, 1995; Krannich et al., 1993, pp. 263–287; National Research Council, 1984). For example, in their study of six rural communities in southern Nevada located near the proposed Yucca Mountain site, Krannich et al. (1993) identified smaller population sizes, relative remoteness, and limited community resources as contributing to rural residents' heightened perceptions of risk.

Concerning energy development and export attitudes broadly, more recent research has examined the effects of "culture regions" (Woodard, 2012)—areas which "transgress political boundaries and are instead characterized by cultural traits, such as religion, language, and history" (Hazboun, 2019). Scholars have evaluated regional attitudes toward coal by rail exports (McGrath, 2018), hydrocarbon exports (Hazboun, 2019; Pierce et al., 2018), as well as natural gas, oil, and coal development (Hazboun & Boudet, 2020), and find mixed evidence to support a unique Pacific Northwest culture region—a region with aggressive renewable portfolio standards and related climate, energy, and environmental policies where numerous hydrocarbon export terminal sitings have been foiled due to civic and policy efforts (Hazboun, 2019). In this region, researchers find nascent evidence that policy, civic, and social contexts may influence greater opposition toward hydrocarbon development and exports relative to other parts of the country.

Relatedly, Zanocco et al. (2020) find evidence to support varied "geographies of perception" vis-à-vis energy development, such that political ideology, proximity, and place can exhibit different effects in different regions. For example, they find that the influence of political ideology—which has been found to influence attitudes toward oil and gas development at different spatial distances—can also vary by both area context and analytic scale, such that support or opposition to energy development may also be influenced by other temporal, hypothetical, or social dimensions (Zanocco et al., 2020).

Other rail transport studies have focused on impacts and perceptions of coal (Lazo & McClain, 1996; McGrath, 2018) and toxic chemicals (Bowler et al., 1994), with limited scholarly attention to crude oil or other liquid hydrocarbons, and notably so in recent years amidst significant increases in OBR exports and activity. As a separate but related study, I employed content analysis of OBR news media coverage and found much of it favored metropolitan areas and issues—such as the potential for catastrophic damage to urban infrastructure and economic sectors, or calls to reduce

OBR speed limits for trains traversing major population centers—with much less consideration for rural areas (Junod, 2020). Finally, and in further contrast with other research examining perceptions of high consequence energy export infrastructure such as nuclear or pipeline siting, no study has been identified which examines and compares attitudes toward OBR prospectively—that is, prior to an accident event.

Methods and analysis

A news media content analysis of OBR activity and impacts (Junod, 2020) guided the purposive selection of four communities for comparative analysis: Mosier, Oregon; Ashville, Ohio; Spokane, Washington; and Columbus, Ohio (Fig. 13.2). Communities were selected for their varying population sizes, geographies, and contexts of OBR impacts. Key informant interviews were conducted with community members from market, state, and civil society sectors (N = 58), which informed the development of a survey instrument to quantitatively explore relationships and variations regarding preference, export, and risk attitudes to OBR and other energy and infrastructure regimes (N = 571). Analyses presented here are drawn from survey findings.

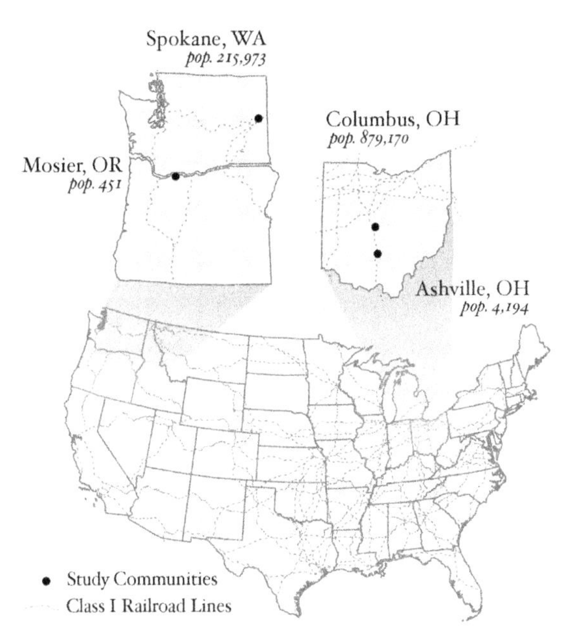

Figure 13.2 Geographical dispersion and location of study communities on Class I rail lines.

Survey dissemination

The survey was deployed throughout 2019 to 1450 randomly sampled residences located within the 1-mile PHMSA-determined impact radius for hazardous freight (United States Department of Transportation Pipeline and Hazardous Materials Safety Administration, 2016). The survey was disseminated via the Drop off/Pick up (DOPU) method, whereby paper surveys are physically dropped off and picked up from sampled residences. Based in social exchange theory, the theoretical premise of the DOPU method is that face-to-face contact and in-person solicitation increases response and completion rates (Dillman et al., 2014). Accordingly, target households were visited repeatedly until contact was made with a resident and solicitation could be made in person. If contact was attempted at a target household but was unsuccessful because a resident did not answer the door or was not home, repeated attempts were made at different times later in the day or on subsequent days until contact was either made or determined to be unsuccessful. Although more time intensive, DOPU has also been shown to be more cost-effective and often yields higher response rates than mail surveys, thus reducing issues associated with nonresponse bias (Jackson-Smith et al., 2016; Singer, 2006; Trentelman et al., 2016).

Participation ranged by community. Overall, complete responses were collected from 571 residences and contact was unsuccessful at 411 residences, representing a response rate of 55% among households where contact was successful and 39% among all households. Responses were lowest in Ashville, Ohio—the rural community with no history of oil train derailment (47% among successful contacts); followed by Columbus, Ohio (48% among successful contacts) and Spokane, Washington (55% among successful contacts)—and highest in Mosier, Oregon, the rural community which experienced a significant OBR derailment in 2016 (75% among successful contacts).

Study communities

Ashville, Ohio is an exurban agricultural village located on the edge of the Appalachian foothills. Key regional agricultural products are corn and soybeans. The municipal boundary is entirely within the 1-mile PHMSA-determined potential impact radius for train accidents involving hazardous freight (PHMSA, 2016). Two major rail lines (CSX and Norfolk Southern) bisect Ashville from north to south, with five railroad crossings as at-grade junctions and two as above-grade trusseled overpasses. No crude oil or other liquid hydrocarbon derailments have occurred in the area.

Columbus, Ohio, the state capital, is located in the center of the state. Columbus is a regional hub for CSX and Norfolk Southern Railroads, the primary railroads serving the eastern Midwest, South, and east coast of the United States. Rail lines bisect Columbus from east to west and north to south, pass via tunnels beneath the downtown core, and run adjacent to interstate and county highways, critical waterways, and dozens of city neighborhoods.

On July 11, 2012, a broken rail track caused 17 cars of a 98-car Norfolk Southern train hauling ethanol and agricultural products to derail just north of downtown (National Transportation Safety Board, 2014). At least one ethanol car was punctured, causing a fire which spread to two other ethanol cars and precipitated a sizable eruption (NTSB, 2014). Two people were injured, approximately 100 nearby residences were evacuated, and infrastructure and other damages were estimated at $1.2 million (National Transportation Safety Board, 2014).

Spokane, Washington is the northwestern hub for Union Pacific and Burlington Northern Santa Fe (BNSF) Railroads, the primary railroads serving the western United States. All crude oil and coal rail exports originating in the Bakken Shale play and Powder River Basin that are destined for Pacific Northwest terminals pass through Spokane. Rail lines traverse the downtown on elevated tracks ranging in height from approximately 20 feet through the downtown core to 175 feet over the Latah Creek canyon directly adjacent. Numerous freight derailments have occurred in the Spokane area, but no crude oil or other liquid hydrocarbon accidents have been reported.

When Mosier, Oregon located approximately 280 miles (450 kilometers) from Spokane experienced a significant OBR accident in 2016, the Spokane City Council voted to make rail transportation of oil and uncovered coal through city limits a civil infraction carrying a $261 per-railcar fine. Although withdrawn due to threatened legal battles, it was later brought up as a ballot initiative and received the support of the Spokane Firefighters Union and numerous faith communities and local advocacy groups (Hill, 2017). BNSF and Union Pacific, western coal mining companies, as well as stakeholders in the then-proposed Vancouver Energy Terminal spent hundreds of thousands of dollars in local media buys, door-to-door leafleting, and opposition funding ahead of the vote, which ultimately failed (Hill, 2017).

Mosier, Oregon is a rural agricultural community located on the southern bank of the Columbia River in the Columbia River Gorge National Scenic

Area. Mosier is an area orcharding hub and popular destination for cyclists and wind surfers. Union Pacific operates a rail line situated along the Columbia River banks that separates Mosier from the river. On the Washington side, BNSF also operates a rail line. All crude oil and coal rail exports that originate in the Bakken Shale and Powder River Basin destined for Pacific Northwest terminals pass through the Columbia River Gorge on either the Union Pacific/Oregon side or the BNSF/Washington side.

On June 3, 2016, 16 cars from a 96-car unit train transporting Bakken crude derailed in Mosier. The Federal Rail Administration later determined the cause of the accident to be numerous broken lag bolts, which were not identified prior to the accident due to Union Pacific Railroad's failure to maintain its track equipment (Federal Railroad Administration, 2016). At least three tanker cars caught on fire and oil from four others spilled into the surrounding soil and community water treatment plant. A number of tank car eruptions—or BLEVEs (boiling liquid expanding vapor explosions)—occurred, spreading the fire into the nearby area. The community school was evacuated, as were approximately one-third of town residents, including all those living in the mobile home neighborhood closest to the derailment. Evacuated residents were unable to return home for 2–3 days and wastewater service was interrupted for over 3 weeks. An estimated 47,000 gallons of oil spilled, 16,000 gallons vaporized or burned, and 2960 tons of irremediable soil were later excavated and removed (Environmental Protection Agency, 2016).

Analysis

As no preliminary or baseline data exist concerning perceptions toward OBR activity and exports in US rail communities, this chapter presents comparative descriptive and significance findings across and between study communities on three key perception questions.

Risk perceptions on a range of potential OBR impacts

Respondents were presented with a list of 17 potential impacts from OBR activity identified through community stakeholder interviews and news media content analysis and asked to indicate their level of concern (No to Low vs. Moderate to High) (Fig. 13.3).

A factor analysis using the principal components method and varimax factor rotation was conducted on these items to assess underlying

	Obs	% No to Low Concern	% Moderate to High Concern
Increased vehicle traffic	559	71.20	28.80
Increased agricultural producer competition for rail space	537	69.09	30.91
Harm to private property rights	555	60.00	40.00
Increased railroad traffic	554	58.48	41.52
Harm to tribal rights or resources	552	53.62	46.38
Increased noise	541	52.87	47.13
Air pollution	561	47.06	52.94
Infrastructure loss or damage associated with a derailment	560	44.29	55.71
General pollution	560	41.79	58.21
Climate change	558	41.40	58.60
Water pollution	556	40.83	59.17
Public injury or death	562	40.39	59.61
Fire associated with a derailment	562	39.68	60.32
Fish or wildlife habitat contamination	561	39.22	60.78
Train derailment	563	38.90	61.10
Explosion associated with a derailment	563	37.83	62.17
Oil spill or leak	561	31.91	68.09

Figure 13.3 Rail export risk objects or outcomes and levels of concern.

dimensions of risk.[1] Factor analysis revealed a three-factor structure, with three-factor components with eigenvalues greater than 1.0. The first factor, comprised of "Catastrophic" items, explained 36% of the variance with six items loading: Oil spill or leak, Train derailment, Infrastructure loss or damage, Fire, Explosion, and Public injury or death. These items exhibited very satisfactory internal consistency (Cronbach's $\alpha = 0.9728$). The second factor, "Natural Resources/Environment," explained 32% of the variance with six items loading onto the factor: Air pollution, Water pollution, Fish or wildlife habitat contamination, General pollution, Harm to tribal rights or resources, and Climate change. These items were also highly correlated (Cronbach's $\alpha = 0.9576$). The third factor, "Day-to-Day Disruption/

[1] A Bartlett test of sphericity was found significant ($P < .001$), indicating items were adequately correlated to run a factor analysis. Next, a Kaiser—Meyer—Olkin (KMO) test of sampling adequacy was conducted to assess the appropriateness of item correlation for factor analysis, such that the overlap between items is sufficient, but not excessive to produce spurious results. A KMO coefficient of 0.5 and higher indicates appropriate correlation, with higher values better; KMO for the risk items was 0.952, indicating excellent sampling correlation for factor analysis.

Nuisance" items, explained 16% of the variance with three items loading: Increased vehicle traffic, Increased railroad traffic, and Increased noise. These were also highly correlated (Cronbach's $\alpha = 0.8498$). Two items did not factor: "Increased agricultural producer competition for railcar space" and "Harm to private property rights." A Pearson chi-squared statistic was calculated to test for significant differences on these items by community and rurality/urbanity.

Community capacity and vulnerability perceptions

Respondents were next asked to share perceptions of community capacity and potential vulnerabilities in the event of an OBR accident (Fig. 13.4).

Respondents indicated whether they were confident in their community's ability to respond to an OBR accident, and if they did not choose "confident," were asked to identify capacity or vulnerability factors that reduced their confidence, identified through community stakeholder interviews and news media content analysis: Limitations of community tax base; Limitations of community emergency funding; Area natural resource or agricultural vulnerabilities; Limitations of industry responsibility to remediate accidents; Rail-adjacent infrastructure vulnerabilities; and Rail-adjacent economic and business sector vulnerabilities. A Pearson chi-squared statistic was calculated to test for significant differences on these items by community and rurality/urbanity.

Overall support and opposition

Finally, respondents were asked to indicate their overall support or opposition to OBR exports traversing their communities. A Pearson chi-squared statistic was calculated to test for significant differences in support and opposition by community and rurality/urbanity.

	Confident	Not Confident
	freq. (%)	freq. (%)
Overall	125 (.23)	407 (.77)
Ashville, OH	13 (.15)	75 (.85)
Columbus, OH	48 (.38)	80 (.62)
Spokane, WA	50 (.26)	144 (.74)
Mosier, OR	14 (.11)	108 (.89)

Figure 13.4 Differences in confidence by study community.

Findings

Risk perceptions

When risk perceptions were compared by rurality and urbanity (i.e., Mosier, OR and Ashville, OH compared to Spokane, WA and Columbus, OH), few similarities and a number of significant differences present (Fig. 13.5).

Across all communities, the nonfactoring and Day-to-Day concerns sorted lowest; however, increased Railroad Traffic and Noise sorted both higher and significantly greater with rural respondents than urban respondents ($P < .001$ and $P < .05$, respectively). Likewise, Catastrophic concern items sorted highest with rural respondents, whereas urban respondents ranked most Catastrophic items across the middle ranges—with the exception Oil Spill, which ranked as most concerning with over 70% indicating moderate-high concern.

Even so, the majority of highest-ranking concerns with urban respondents were Environmental, in contrast to the rural communities where Environmental items tended to sort across the middle and lower ranges. One Environmental item, Climate Change, sorted much higher and was of significantly greater concern in the urban communities than the rural communities ($P < .05$).

Community capacities and vulnerabilities

Perceptions of community capacity and vulnerability varied by community, with urban respondents indicating lower levels of concern than rural respondents on most items (Fig. 13.6).

No significant differences presented between urban and rural respondents on infrastructure and economic/business sector vulnerabilities; however,

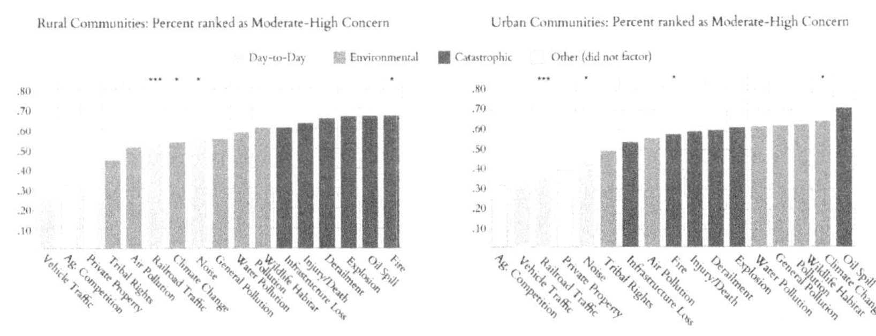

Figure 13.5 Differences in risk objects or outcomes of concern by rural and urban communities.

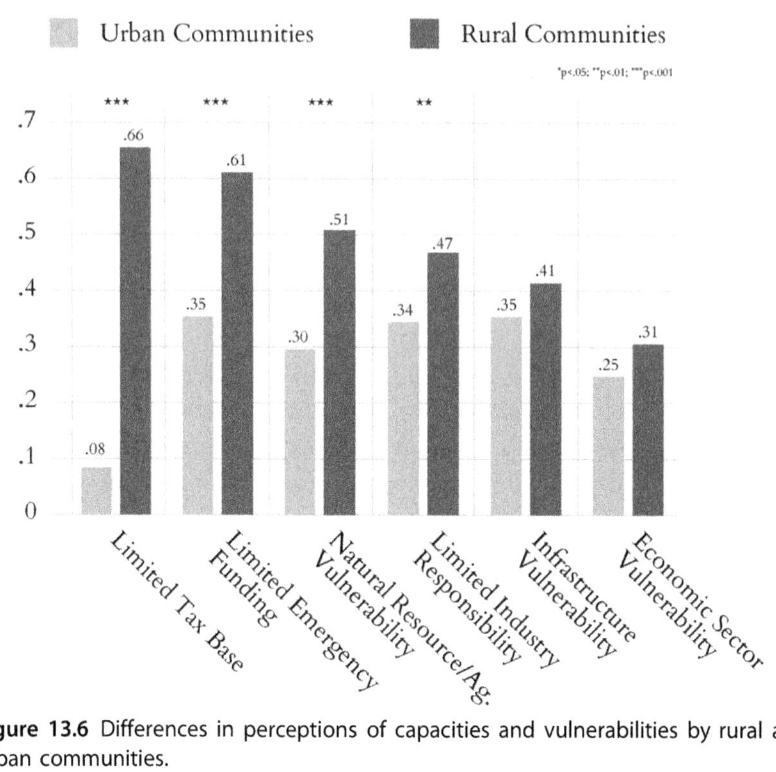

Figure 13.6 Differences in perceptions of capacities and vulnerabilities by rural and urban communities.

rural respondents were significantly likelier to consider small community tax bases and emergency funding resources as well as nearby natural or agricultural resource vulnerabilities as limiting factors to their community's capacity to respond to an OBR accident ($P < .001$). Likewise, rural respondents were likelier than urban respondents to cite perceived limitations in industry responsibility to address accidents as contributing to community vulnerability ($P < .01$).

Support, opposition, and uncertainty

Across the study communities, residents expressed slightly greater opposition than support (Fig. 13.7). Examined by community, however, respondents exhibit a stepwise increase in opposition and decrease in support, with Ashville respondents expressing the highest support (49%), followed by Columbus (39%), Spokane (34%), and Mosier (24%). Likewise, uncertainty toward OBR exports was greatest in Columbus and Ashville, lower in Spokane, and least in Mosier. A pairwise test for variation in support versus opposition by community was significant on all comparisons ($P < .05$), but rurality/urbanity was not.

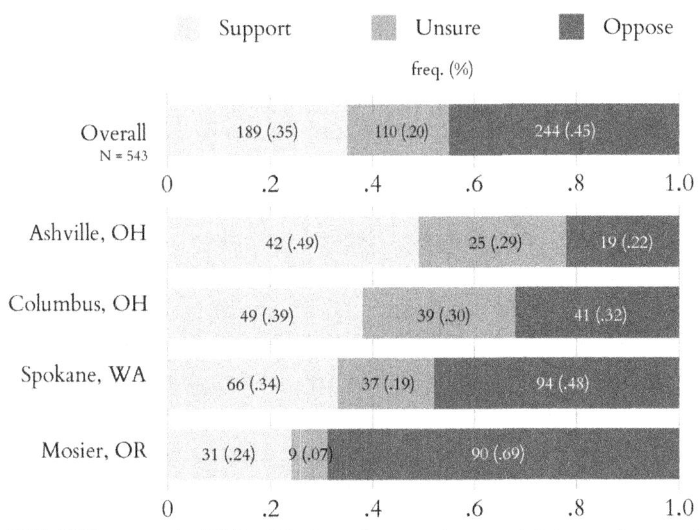

Figure 13.7 Differences in OBR attitudes of support, uncertainty, and opposition by study community.

Discussion: rural and urban risks, vulnerabilities, and opposition

Public perceptions toward OBR activity and exports have received limited social scientific attention in environmental social science and energy development literatures. Three key findings emerge from this study which reduce this gap and signal directions for further study.

First, I identify key risk domains associated with OBR about which community perceptions vary: Day-to-Day risks like increased traffic and noise, Environmental risks like climate change and pollution, and Catastrophic risks like fire, explosion, and injury or death. Controlling for nothing else, significant differences exist between rural and urban respondents on perceptions of these risks, with urban respondents exhibiting greater concern for Environmental issues like climate change, and rural respondents exhibiting greater concern for both Catastrophic and Day-to-Day impacts, like fire, noise, and traffic.

Lower concern for climate change in rural communities is consistent with findings from other environmental attitudes studies (Hazboun & Boudet, 2020; Whitmarsh, 2011). There is nothing about rurality per se that influences lower climate change belief or concern; rather, other individual and experiential predictors like political ideology—which tends to be the most durable influence—in addition mixed effects of age, educational attainment, culture, and in-group influences are found to contribute to variations in climate change views (Kahan et al., 2007; Hazboun & Howe, 2018; Hazboun & Boudet, 2020).

Although not directly examined here, differences in community *construal* of OBR may also influence qualitative differences in perceived risks, resulting in greater concern for more abstract impacts like climate change in urban communities and greater concern for more concrete impacts like traffic or fire in rural communities. The theoretical premise of Construal Level Theory (CLT) is that individuals' psychological distance from a phenomenon influences their perception of it (Trope & Liberman, 2010). CLT suggests that phenomena which are psychologically closer—in time, distance, possibility, and relationally—are construed with greater detail and concreteness than phenomena which are further away, which in turn tend to be more abstracted (Spence et al., 2012). Other studies have found construal effects on attitudes toward hydraulic fracturing (Clarke et al., 2016; Zanocco et al., 2020) and climate change (Spence et al., 2012).

In the rural communities examined here, railroads are a dominant dimension of community infrastructure relative to the urban communities, where daily interaction with OBR is less intimate given the relative scale of urban infrastructure, development, and sprawl. In the smaller rural communities—both located entirely within a 1-mile radius of rail infrastructure—residents interact with the railroad with greater frequency, which may associate with greater concreteness in their estimation of potential impacts compared with urban respondents, who may construe impacts with greater psychological distance. As a result, it is possible urban respondents may not perceive OBR risks as concretely as rural residents, as is found by the greater relative concern for climate change—a more abstract risk (Spence et al., 2012)—vis-à-vis more concrete risks, such as noise, traffic, or fire, in rural communities. Even so, Zanocco et al. (2020) find that proximity or exposure alone do not consistently explain variation in energy views, and that place, proximity, and ideology can interact differently in different regions. These interactions and their potential geographic or regional differences are apt for deeper analysis in the context of OBR.

I also find significant differences in perceptions of vulnerability and capacity, with rural respondents identifying many of the same vulnerabilities—in addition to multiple others—as those which are commonly attributed to urban areas. In line with the limited literature identifying rural concerns regarding other high-consequence energy regimes (Krannich & Albrecht, 1995; Krannich et al., 1993, pp. 263–287;

National Research Council, 1984), rural respondents examined here perceive lower capacity and higher vulnerability concerning potential OBR impacts than do urban respondents. At the same time, there is no substantive difference between rural and urban respondents on built infrastructure or economic and business sector vulnerabilities. Indeed, rural respondents identify these concerns *as well as* other vulnerabilities associated with their relatively small size, area natural resources, and distance from urban areas. That is, rural respondents perceived similar levels of risk as urban respondents concerning infrastructure and economic sector vulnerabilities, in addition to numerous other more rural-specific concerns which are identified much less in urban areas.

This stands in contrast with news media reporting on OBR risks and impacts, which often frames infrastructure and economic sector vulnerabilities to OBR as urban issues (Junod, 2020), and portends diminished or metro-biased public understandings of potential OBR impacts, as news media presents key sources of public information concerning energy development and influences public views (Ashmoore et al., 2016). This gap may have serious implications for state and municipal emergency planning and resource allocation in rural areas given the role of media in shaping public views.

Finally, I find that greater perceived risks, vulnerabilities, and capacity limitations do not consistently translate into opposition, as is evident by the contrast between Mosier respondents' opposition to OBR and Ashville respondents' support despite similar risk and vulnerability perceptions. Hazboun (2019) and Hazboun & Boudet (2020) have identified emerging evidence for "culture regions" which may influence public attitudes toward energy regimes and related infrastructure, and Zanocco et al. (2020) identify the influence of context and experience in addition to political ideology and proximity in shaping public views. Particularly outside the Pacific Northwest, the presence, shape, and influence of such regions remain underexamined and are ripe for middle-range empirical testing (Zanocco et al., 2020). Even so, there are parallels to other energy development social science which has examined the colonizing effects of single-industry extractive activity—such as coal mining—whereby industry actors cultivate regional identities tied to extractive activity and thereby secure social license and public support despite negative environmental, economic, and public health externalities (Bell & York, 2010; Junod, 2020; Junod & Jacquet, 2019).

In comparing Ashville to Mosier, the two rural communities—one which experienced a significant derailment and is located in the Pacific Northwest, and one which has not and is located on the edge of Appalachian coal country—it is possible that although Ashville respondents identify concrete risks associated with greater psychological proximity to OBR, the fact that they have not experienced an accident makes a potential future OBR problem seem "further away." Furthermore, in a hydrocarbon production state like Ohio—where coal consumption and natural gas production rank among the top five in the nation, where wind developers face among the strictest siting and setback requirements in the country, and the energy production portfolio is over 88% carbon-based—the cultural and economic contexts of oil, gas, and coal production may attenuate otherwise heightened risk perceptions and influence lower levels of OBR opposition. This contrasts with the Pacific Northwest, where hydrocarbon production has never ascended to the levels of the Appalachian region, and where other studies and this research have found stronger opposition to hydrocarbon activity.

This research presents a first step in understanding the risk perceptions, vulnerabilities, and OBR export attitudes in places experiencing the downstream effects of 21st century shale energy revolution. This study does not and was not designed to examine important individual-level predictors which may also influence these perceptions and views, but rather presents baseline comparisons between different sized communities, in different regions, with different experiences with OBR. This study also does not include sophisticated analyses of potential mediating or path effects between relevant predictor variables and attitudes; this second step of analysis is in preparation.

Conclusion

In the context of energy exports, the variegated "impact geographies" (Haggerty et al., 2018) or "geographies of perception" (Zanocco et al., 2020) of 21st century energy expansion are only nascently understood. Impacts from expanding exports will not be felt evenly, with a variety of community, policy, and social factors influencing both public perceptions and community and environmental outcomes. Understanding public perceptions of different energy export regimes in different regions is vital for state-level energy development planning, yet little was previously known about public perceptions of OBR.

As many communities across the United States have moved to limit potential risks associated with oil train activity, this research presents an important first step in understanding community-level differences in risk, vulnerability, and support versus opposition perceptions. Metropolitan-centered reporting and risk mitigation planning may miss rural agricultural and natural resource vulnerabilities, as well as the potential that rural communities with small tax bases or those in states with significant devolution of emergency management and funding may be at greater relative risk to OBR accidents than better resourced urban areas. This study also supports and extends the nascent culture regions and geography of perceptions literatures in identifying regional variation between Pacific Northwest and Midwest/Appalachian support for OBR, such that the influence of context and experience may also shape public views.

In so doing, this research internalizes rural contexts to the issue of OBR transportation and exports and may support future community supported policy which reflects the unique concerns of rural and metropolitan communities alike.

References

American Association of Railroads. (2017). *U.S. Rail crude oil traffic.* https://www.aar.org/data/u-s-rail-crude-oil-traffic/.

Ashmoore, O., Evensen, D., Clarke, C., Krakower, J., & Simon, J. (2016). Regional newspaper coverage of shale gas development across Ohio, New York, and Pennsylvania: Similarities, differences, and lessons. *Energy Research and Social Science, 11*, 119–132. https://doi.org/10.1016/j.erss.2015.09.005

Bell, S. E., & York, R. (2010). Community economic identity: The coal industry and ideology construction in West Virginia. *Rural Sociology, 75*(1), 111–143. https://doi.org/10.1111/j.1549-0831.2009.00004.x

Boudet, H. S., Jayasundera, D. C., & Davis, J. (2011). Drivers of conflict in developing country infrastructure projects: Experience from the water and pipeline sectors. *Journal of Construction Engineering and Management, 137*(7), 498–511. https://doi.org/10.1061/(ASCE)CO.1943-7862.0000333

Bowler, R. M., Mergler, D., Huel, G., & Cone, J. E. (1994). Psychological, psychosocial, and psychophysiological sequelae in a community affected by a railroad chemical disaster. *Journal of Traumatic Stress,* 7(4), 601–624. https://doi.org/10.1007/BF02103010

Brisson, G., & Bouchard-Bastien, E. (2018). With or without railway? *Post-catastrophe Perceptions of Risk and Development*, 123–136.

Clarke, C. E., Budgen, D., Hart, P. S., Stedman, R. C., Jacquet, J. B., Evensen, D. T. N., & Boudet, H. S. (2016). How geographic distance and political ideology interact to influence public perception of unconventional oil/natural gas development. *Energy Policy, 97*, 301–309. https://doi.org/10.1016/j.enpol.2016.07.032

Clay, K., Jha, A., Muller, N., & Walsh, R. (2017). *The external costs of transporting petroleum products by pipelines and rail: Evidence from shipments of crude oil from North Dakota*. National Bureau of Economic Research.

Dante, B. J., Lucio, T. F. G., & Russo, S. M. R. (2016). Social, technical-economic, environmental and political assessment for the evaluation of transport modes for petroleum products. *Journal of Transport Literature*, 25–29. https://doi.org/10.1590/2238-1031.jtl.v10n1a5

Dillman, D., Smyth, J., & Christian, L. (2014). *Internet, phone, mail, and mixed-mode surveys: The tailored design method*.

Emmanuelle, B.-B., & Geneviève, B. (2018). Entre attachement aux lieux et gestion de la reconstruction post-sinistre : l'action municipale au centre-ville de Lac-Mégantic, Québec (Canada). *Norois*, 75–88. https://doi.org/10.4000/norois.7201

Environmental Protection Agency. (2016). *Mosier Oregon train derailment*. https://response.epa.gov/site/site_profile.aspx?site_id=11637.

Federal Railroad Administration. (2016). *Accident investigation report HQ-2016-1136*. https://railroads.dot.gov/sites/fra.dot.gov/files/fra_net/17931/HQ-2016-1136%20Final_Mosier%2C%20OR.pdf.

Haggerty, J., Kroepsch, A., Bills Walsh, K., Smith, K., & Bowen, D. (2018). Geographies of impact and the impacts of geography: Unconventional oil and gas in the American West. *The Extractive Industries and Society, 5*(4), 619–633.

Hazboun, S. O., & Boudet, H. S. (2020). Public preferences in a shifting energy future: Comparing public views of eight energy sources in north America's Pacific Northwest. *Energies, MDPI, Open Access Journal, 13*(8), 22. https://ideas.repec.org/a/gam/jeners/v13y2020i8p1940-d345777.html.

Hazboun, S. O., & Howe, P. (2018). Public opinion on climate change in rural America: A potential barrier to resistance. In P. Lachapelle, & D. Albrecht (Eds.), *Addressing climate change at the community level in the United States*. Routledge.

Hazboun, S. O. (2019). A left coast 'thin green line'? Determinants of public attitudes toward fossil fuel export in the northwestern United States. *Extractive Industries and Society, 6*(4), 1340–1349. https://doi.org/10.1016/j.exis.2019.10.009

Heikkila, T., Pierce, J. J., Gallaher, S., Kagan, J., Crow, D. A., & Weible, C. M. (2014). Understanding a period of policy change: The case of hydraulic fracturing disclosure policy in Colorado. *The Review of Policy Research, 31*(2), 65–87. https://doi.org/10.1111/ropr.12058

Hill, K. (2017). Spokane voters reject fines for coal, oil trains traveling through downtown. *Spokane Spokesman-Review*.

Jackson-Smith, D., Flint, G., Dolan, M., Trentelman, C., Holyoak, G., Thomas, B., & Ma, G. (2016). Effectiveness of the drop-off/pick-up survey methodology in different neighborhood types. *Journal of Rural Social Sciences, 31*(3), 35–67.

Junod, A. (2020). *Risks, attitudes, and discourses in hydrocarbon transportation communities: Oil by rail and the United States' shale energy revolution*.

Junod, A. N., & Jacquet, J. B. (2019). Shale gas in coal country: Testing the Goldilocks Zone of energy impacts in the western Appalachian range. *Energy Research and Social Science, 55*, 155–167. https://doi.org/10.1016/j.erss.2019.04.017

Kahan, D. M., Braman, D., Gastil, J., Slovic, P., & Mertz, C. K. (2007). Culture and identity-protective cognition: Explaining the white-male effect in risk perception. *Journal of Empirical Legal Studies, 4*(3), 465–505. https://doi.org/10.1111/j.1740-1461.2007.00097.x

Krannich, R. S., & Albrecht, S. L. (1995). Opportunity/threat responses to nuclear waste disposal facilities. *Rural Sociology, 60*(3), 435–453. https://doi.org/10.1111/j.1549-0831.1995.tb00582.x

Krannich, R. S., Little, R. I., & Cramer, L. A. (1993). *Rural community residents' views of nuclear waste repository siting in Nevada*. Duke University Press. https://doi.org/10.1215/9780822397731-010

Lazo, J. K., & McClain, K. T. (1996). Community perceptions, environmental impacts, and energy policy. *Energy Policy*, 531–540. https://doi.org/10.1016/0301-4215(96)00025-0

Mark, W. (2018). Justice denied: Why was there no public inquiry into the Lac-Mégantic disaster? *Revue Generale de Droit*, 131–154. https://doi.org/10.7202/1047375ar

McAdam, D., Boudet, H., Davis, J., Orr, R., Scott, W., & Levitt, R. (2010). "Site fights" explaining opposition to pipelines in the developing world. *Political Science and Politics, 28*(1), 57–72.

McGrath, M. (2018). *The power and politics of health impact assessment in the Pacific Northwest coal export debate*.

Mélissa, G., Geneviève, P., Danielle, M., Mathieu, R., Robert, S., Sonia, B., M, S. J., & Linda, P. (2014). The public health response during and after the Lac-Mégantic train derailment tragedy: A case study. *Disaster Health*, 113–120. https://doi.org/10.1080/21665044.2014.1103123

National Research Council. (1984). *Social and economic aspects of radioactive waste disposal: Considerations for institutional management*.

National Transportation Safety Board. (2014). *Railroad accident brief DCA12MR006*. https://www.ntsb.gov/investigations/AccidentReports/Reports/RAB1408.pdf.

Pierce, J. J., Boudet, H., Zanocco, C., & Hillyard, M. (2018). Analyzing the factors that influence U.S. public support for exporting natural gas. *Energy Policy, 120*, 666–674. https://doi.org/10.1016/j.enpol.2018.05.066

Rabe, B. G., & Borick, C. (2013). Conventional politics for unconventional drilling? Lessons from Pennsylvania's early move into fracking policy development. *The Review of Policy Research, 30*(3), 321–340. https://doi.org/10.1111/ropr.12018

Santiago-Martín, Guesdon, G., Díaz-Sanz, J., & Galvez-Cloutier, R. (2015). Oil spill in Lac-Mégantic: Environmental monitoring and remediation. *International Journal of Water and Wastewater Treatment, 2*(1).

Singer, E. (2006). Introduction: Nonresponse bias in household surveys. *Public Opinion Quarterly, 70*(5), 637–645. https://doi.org/10.1093/poq/nfl034

Spence, A., Poortinga, W., & Pidgeon, N. (2012). The psychological distance of climate change. *Risk Analysis, 32*(6), 957–972. https://doi.org/10.1111/j.1539-6924.2011.01695.x

Trang, T., Taylor, L. C., Boudet, H. S., Keith, B., & Peterson, H. L. (2019). Using concepts from the study of social movements to understand community response to liquefied natural gas development in clatsop county, Oregon. *Case Studies in the Environment*, 1–7. https://doi.org/10.1525/cse.2018.001800

Trentelman, C., Irwin, J., Petersen, K., Ruiz, N., & Szalay, C. (2016). The case for personal interaction: Drop-off/pick-up methodology for survey research. *Journal of Rural Social Sciences, 31*(3).

Trope, Y., & Liberman, N. (2010). Construal-level theory of psychological distance. *Psychological Review, 117*(2), 440–463. https://doi.org/10.1037/a0018963

United States Department of Transportation Pipeline and Hazardous Materials Safety Administration. (2014). *Draft regulatory impact analysis [docket No. PHMSA-2012-0082] (HM-251) hazardous materials: Enhanced tank car standards and operational controls for high-hazard flammable trains*. Notice of Proposed Rulemaking. https://www.phmsa.dot.gov/hazmat/regs/rulemaking/nprm-anprm.

United States Department of Transportation Pipeline and Hazardous Materials Safety Administration. (2016). *Emergency response guidebook*. https://doh.sd.gov/documents/EMS/EmergRespGuidebook.pdf.

United States Energy Information Administration. (2020). *EIA expects U.S. net natural gas exports to almost double by 2021.* https://www.eia.gov/todayinenergy/detail.php?id=42575.

Whitmarsh, L. (2011). Scepticism and uncertainty about climate change: Dimensions, determinants and change over time. *Global Environmental Change, 21*(2), 690–700. https://doi.org/10.1016/j.gloenvcha.2011.01.016

Woodard, C. (2012). *American nations: A history of the eleven rival regional cultures of north America.*

Zanocco, C., Boudet, H., Clarke, C. E., Stedman, R., & Evensen, D. (2020). NIMBY, YIMBY, or something else? Geographies of public perceptions of shale gas development in the Marcellus shale. *Environmental Research Letters, 15*(7). https://doi.org/10.1088/1748-9326/ab7d01

CHAPTER 14

Leave it in the ground, or send it abroad? Assessing themes in community response to coal export proposals using topic modeling of local news

Shawn Hazboun[1], Kathleen Saul[1], Huy Nguyen[2] and Richard Weiss[2]
[1]Graduate Program on the Environment, The Evergreen State College, Olympia, WA, United States;
[2]Computer Sciences Program, The Evergreen State College, Olympia, WA, United States

Introduction

Since the start of the recession of 2008, the demand for and production of coal in the United States has been declining as cheap natural gas has replaced coal-fired electrical generation; simultaneously, renewable energy use has expanded (Fig. 14.1). Despite ample supply and domestic and international markets for the product, coal exports accounted for just over 9% of production in 2014 and only 13.75% of production in 2019 (Energy Information Administration, 2021). The US Energy Information Administration (EIA) has predicted that coal production in the United States will remain flat or continue to decline in the years ahead (Energy Information Administration, 2017). US coal companies have turned to the export market to make up for deficits in domestic sales, shipping to countries like India, Brazil, Japan, and South Korea that rely on imported coal to fire power plants and support their expanding economies (Energy Information Administration, 2021). Yet, producers in the United States have had trouble expanding transportation routes for export, which most often involves developing lines of rail transport across multiple states as well as building a new export terminal or expanding capacity at an existing export site.

Currently, coal is shipped from interior Western States by train through the Seattle Customs District, then another 150 miles to export terminals in Vancouver, British Columbia for shipment overseas (Energy Information

Public Responses to Fossil Fuel Export
ISBN 978-0-12-824046-5
https://doi.org/10.1016/B978-0-12-824046-5.00006-0

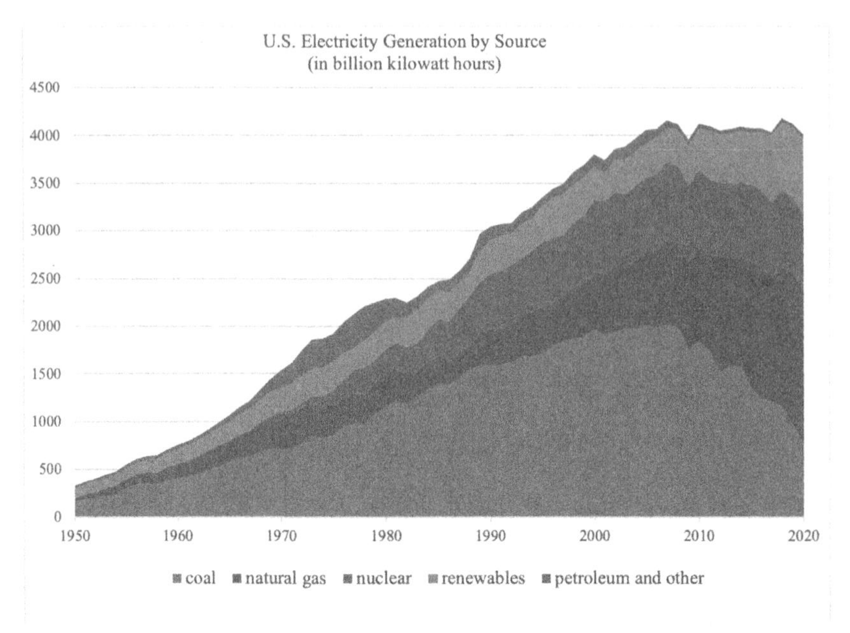

Figure 14.1 US electricity generation by major energy source, 1950–2020. (Energy Information Administration, 2021).

Administration, 2018). Coal companies have long been pursuing the development of greater West Coast coal export capacity. The deep-water port towns of the West Coast are desirable to Western coal producers as sites of expansion for export facilities. The towns could provide the nearest and most direct export route from coal seams to ocean shipping channels, especially for producers in the Powder River Basin of Wyoming and Montana, yet limited facilities currently form a bottle neck for coal export. Expansion efforts also have ignited intense controversy among stakeholders in this fairly politically progressive and environmentally conscious region. Starting in 2010, six proposals for major coal export terminals were advanced in the states of Washington and Oregon, to which shipping by rail would be more competitive in pricing than to terminals in California (U.S. Army Corps of Engineers, 2016). However, all have been either denied by state permitting agencies or abandoned due to significant public protest.

This chapter examines community responses to two high-profile coal export proposals in Washington State through analysis of local news coverage and letters to the editor over a 10-year time period. The two proposals—called the Gateway Pacific Terminal (GPT) proposal and the

Millennium Bulk Terminals (MBT) proposal—were similar in scope and generated both ample opposition and support within the towns of Bellingham and Longview, respectively. Yet, each proposal was ultimately denied. We examine key themes in local response that emerged across the two cases. We utilize computer-assisted topic modeling, a method increasingly used to conduct content analysis of large quantities of text. We draw on social movement studies and the concept of movement "scale shift" to assess what proportion of contention over coal export terminals region is "scaled up" as a movement with regional awareness, rather than two isolated instances of local contention.

Background and framework

Energy policy and fossil fuel export in Western Washington State

Washington State has been a leader in recent years in domestic climate and energy policy and the transition from coal and other fossil fuels toward non-carbon emitting electricity sources. Coal-fired generation has been almost entirely phased out in Washington State and neighboring Oregon State. Washington's last coal plant, the Centralia Coal Plant owned by TransAlta of Canada, shut down its first burner at the end of 2020, and aims to decommission the remaining burner by the end of 2025. Furthermore, Washington's 2019 Clean Energy Transformation Act (CETA) required utilities providing electricity to Washington residents to fully eliminate greenhouse gas–emitting energy sources from their state portfolios by 2045. CETA puts Washington State among 13 states in the United States to enact a 100% clean energy or renewable energy target (though some of these states only have the target as a "goal," such as Nevada and Wisconsin, rather than as a mandate or law). It also puts pressure on local utilities, like Puget Sound Energy, to plan for a just transition away from ownership of coal-fired power plants elsewhere and the "coal by wire" that still makes up 35% of their electricity resource mix. Such a transition engages local community stakeholders to ensure environmental concerns have been addressed and good, family wage jobs are created for those losing theirs in the move to renewable energy systems.

Despite the state's progress on eliminating coal from its electricity portfolio, Washington remains engaged in the fossil fuel industry through refining and shipping. Washington hosts five major crude oil refineries supplied by a network of pipelines and oil trains, comprising the fifth greatest capacity of any US state to refine oil (Energy Information Administration, 2019). Additionally, several communities in Washington have considered proposals for some of the most highly contentious fossil fuel export facilities nationwide (McClure, 2021; Sightline Institute, 2018; Williams-Derry & de Place, 2017); two are the focus of this chapter.

"Scaled-up" social movement activities: from locally unwanted land uses to broader contention

The major proposals for fossil fuel export facilities have led to a significant amount of news coverage and public debate and contention over fossil fuel transport and export throughout Washington and the broader Pacific Northwest region (McClure, 2021). Regional opposition activity to block new fossil fuel export and transport infrastructure around the region has been dubbed a "thin green line" and is envisioned as a barrier between fossil fuel producers and overseas markets as part of the global effort to reduce carbon dioxide emissions (McClure, 2021; Sightline Institute, 2018; Williams-Derry & de Place, 2017). Opposition actors consist of local citizens, environmental activists, tribal nations, policymakers, and local jurisdictions that have passed resolutions or ordinances to block new fossil fuel transport facilities (Hellegers, 2021b, a; Sightline Institute, 2018; Wysham, 2016). Yet, the proposed export facilities received ample support from local, regional, state, and national actors, many of whom outspokenly criticized Washington Governor Jay Inslee and other regional politicians for being ideological motivated against coal. In a 2018 opinion editorial in *The New York Times*, then Attorney General Fox of Montana argued, "The saga of coal production and export over the last decade is fraught with controversy and competing interests. Politicians in Washington and Oregon have taken extraordinary steps to prevent exports of raw coal and stop the generation of electricity by burning coal ... America simply cannot allow one state to use geography and ideology to discriminate against another state's commodities" (Fox, 2018).

For years, the first two authors have closely followed the debate over the coal export proposals highlighted in this study, as well as broader narratives about fossil fuel export in the region. We observed that several "scaled-up" themes of opposition and support were prominent alongside local concerns about impacts to traffic patterns, local waterways, and health; and we noted that the scaled themes of contention diverged somewhat from classic understanding of how communities respond to "locally unwanted land uses" (Bohon & Humphrey, 2000; Popper, 2010). That is, we observed that regional, national, and global considerations appeared alongside the standard themes of local impacts/risks (such as to the local environment) and benefits (such as creating local jobs) in the debate over whether or not the coal export terminal should be built in each community. One example of such a "scaled" theme is the incorporation of concerns about global carbon emissions employed by actors on both sides of the export terminal debate and at various scales of influence. Local opponents as well as regional and national environmental groups framed the coal terminals as opening a Pandora's box of greenhouse gas emissions by expanding the markets in which US coal could be burned (McClure, 2021; Sightline Institute, 2018; Williams-Derry & de Place, 2017). Conversely, supporters of the terminals argued that expanding exports of US coal to Asian countries like Japan and South Korea would help mitigate global climate change by offsetting the amount of "lower quality" coal these countries import from other producers. Former Attorney General Fox captured this argument succinctly in his opinion editorial, referenced above: "If denying coal ports in Washington is part of the fight against climate change, it is moving the needle in the opposite direction" (Fox, 2018).

As we observed this and other "scaled-up" themes of contention and support, we became curious about the extent to which local concerns were driving the debate over the coal export terminals, compared with regional, national, and global concerns. Another aspect of the same question is to ask to what extent each case portrays a "NIMBY" versus a "NIABY" social response (Boudet, 2011; Dear, 1992). While "NIMBY" (not in my backyard) highlights local opposition to a proposed facility on the basis of concerns about local impacts, "NIABY" (not in anyone's backyard) objections highlight more universal concerns and thus can be deployed by actors outside the local context. Broader opposition movements, such as those waged at regional and national levels against nuclear energy facilities (Hasegawa, 2010, pp. 63–79; Joppke, 1991; Koopmans & Olzak, 2004; Useem & Zald, 1982), employ universal "NIABY" framing of the issue and in doing so make it salient for the broader public living beyond the confounds of the affected community. Snow and Benford label this strategy "frame bridging" (Benford & Snow, 2000; Snow & Benford, 1992). Frame bridging refers to the application of a collective action frame used to generate social movement activity on one issue to another, separate issue. An example of this is utilizing the framing of fossil fuel extraction as a dangerous exacerbator of global climate change (currently used by "leave it in the ground" social movement actors (Erickson et al., 2018; Piggot, 2018; LINGO, 2019) and applying this framing to the debate over expanding export capacity for fossil fuels.

Social movement scholars (Boudet, 2011; Tarrow & McAdam, 2004) refer to the scaling up from local to regional contention as "scale shift" and propose several mechanisms for how social movement actors can scale up local contention over a proposed project to a larger level, thus waging a more influential campaign that is more likely to succeed in opposing the project. These mechanisms include factors like "brokerage" (in which actors connect two or more unrelated sites of contention) and "relational diffusion" (in which interactions, information, and strategies or innovations are diffused across a broader social network).

It is beyond the scope of the present study to answer the question of *how* local social movement activity "scales up" to be part of a broader regional movement. Instead, we address the following research questions: *How do the Millennium Bulk Terminals and Gateway Pacific Terminal cases differ in terms of the themes that were most prominent in community response? To what extent did local concerns drive the debate over the GPT and MBT proposals, compared with geographically broader, "scaled up" concerns?*

Data and methods

We focus on two areas in Washington State where major coal export facilities were proposed—one near the city of Bellingham and one in the city of Longview. We combine case study–style research with computer-assisted topic modeling to understand and compare local and scaled themes of opposition and support in both cases. Case study research proves valuable for close, in-depth understanding of phenomena of interest, including what happened and why or how it happened, as well as the context and characteristics of the phenomena (Gerring, 2004; Yin, 2017). Comparative case studies, in particular, "tend to see the world in terms of people, situations, events, and the processes that connect these" (Bartlett & Vavrus, 2017). Comparing these two cases allows us to understand similarities and differences and the factors that gave rise to them. Examining and comparing themes in two communities' response to coal terminal proposals also helps contextualize and unpack local and "scaled-up" themes in contention over the transportation and export of fossil fuels through

Washington and the greater Pacific Northwest region—similar themes, for example, may suggest a broader, coordinated movement against fossils export across the "thin green line."

Description of the cases: Gateway Pacific Terminal (GPT) and Millennium Bulk Terminals (MBT)

The GPT would have handled almost 60 million tons (US) of dry bulk commodities annually, mostly coal from the Powder River Basin, through northwestern Washington via Cherry Point, just north of the city of Bellingham. The site, located between the BP Cherry Point Refinery (an oil refinery) and the Alcoa Intalco Works aluminum smelter, had already been zoned for industrial use. In early 2011, an international shipping company, SSA Marine, announced interest in constructing an export terminal to ship grain to Asian markets. Drawing on existing permits from 1997, the company began work at the site in the summer of 2011, clearing wetlands and drawing the ire of local residents, environmentalists, and tribal citizens. Project initiation documents from SSA Marine indicated the GPT would not handle grain, but instead process 8.2 million tons of cargo per year, including petroleum coke from the nearby refinery, iron ore, sulfur, potash, and wood chips. Later in 2011, it became clear that SSA Marine was intending to mainly ship coal (Whatcom County, 2011).

The coal export terminal would have resulted in 18 additional train trips per day along the waterfront and through downtown Bellingham to Cherry Point, a route that passes directly through highly popular waterfront Boulevard Park; each train would have been about a mile and a half-mile long (Schultz, 2011). Whatcom County officials ruled that because the coal export terminal far exceeded the intent of the original permits, SSA Marine needed to reapply and meet the more stringent laws for shoreline development. The proposal drew national attention, motivating prominent anticoal activists like Bill McKibben, doctors and health professionals, groups like the Sierra Club, Columbia Riverkeeper, and Washington tribes to oppose the project (Rice, 2011). Although the proponents of the project never submitted their Environmental Impact Assessment, students at Western Washington's Huxley College of the Environment completed one of their own and concluded that the most significant impacts would result from fugitive coal dust from the coal trains, dust that would settle on land, migrate into the waterways, and foul the air all along their route (Brownell et al., 2012). In 2015, the Lummi Nation sent a letter to the US Army Corps of Engineers, asking the agency to immediately reject a permit application for the terminal, arguing that its construction would impinge on their ability to fish in their usual and accustomed areas, thus violating their treaty rights and a 2000 court decision (Schwartz, 2015). Siding with the Lummi Nation, the Army Corps of Engineers denied key permits, and the developer of the terminal withdrew their applications in 2017 (Allen et al., 2017).

The second case, the MBT project, was proposed in early 2012 by developer Lighthouse Resources Inc. in the town of Longview, located in southwest Washington along the Columbia River. Though both located in Western Washington, Bellingham and Longview are quite different in terms of economic activity and socioeconomic indicators. Unlike Bellingham, a college town with a population of just over 90,000 and median household income of $53,396 (U.S. Census Bureau, 2020), Longview is home to about 38,000 people with a median household income of $44,957. Forty four percent of Bellingham residents have a bachelor's degree or higher compared to 16% of those in Longview. Bellingham is considered a politically liberal town, with Whatcom County historically voting blue (60.4% of votes went for Democrat Joe Biden in the 2020 presidential election), while Longview is more politically conservative, with Cowlitz County voting 57.1% in favor of Republican Donald Trump in the 2020 presidential election ("Washington Election Results 2020: Live Results by County," 2021).

The MBT project would have brought an estimated 2650 direct and indirect jobs to the region during the site cleanup and construction of the new facility, and 135 permanent family wage jobs (Richards, 2017). Labor unions and the building trade councils supported the initiative to build what would have been the largest coal export facility in North America. Though there was much support within Longview for the coal terminal, several towns along the rail route expressed concern about the increased train traffic and the impact of fugitive coal dust on air and water pollution (City of Vancouver, 2016). The terminals, rail lines, and docks would have moved over 44 million tons of coal a year to Japan, South Korea, China, and elsewhere. The export terminal would have operated 24 hours per day, 7 days per week, adding 70 vessels per month, or 1680 round trip vessel transits in the Columbia River annually (U.S. Army Corps of Engineers, 2016).

Washington State rejected the US Army Corps of Engineers draft environmental review in 2016 on the basis that it did not address the issues of rail transport of coal through the state and significant environmental impacts like diesel pollution and greenhouse gas emissions (Seattle Times Staff, 2017). In 2017, the Department of Natural Resources denied Millennium's requested sublease on state aquatic lands of the old Reynolds Metals Co./Northwest Alloys dock on the Columbia River, a decision upheld by the Washington Supreme Court in 2020 ("High Court Upholds DNR Denial of Sublease for Longview Coal Dock," 2020). In another blow, the Washington Department of Ecology denied the developer's key water quality permit in 2017. Although MBT pursued legal action, it eventually faced bankruptcy after years of court battles, a steep decline in US coal consumption, and as prices paid for coal on the international market fell (Barrera, 2021). Much public and political contention remains over the state's decisions in this case, as well as regarding the future of the region as a hub for or a blockade against export of US fossil fuels.

Data acquisition and cleaning

We utilized computer-assisted content analysis of local newspaper articles and letters to the editor—for GPT we searched *The Bellingham Herald,* and for MBT we searched *The Longview Daily*. To obtain the data, we performed several keyword searches using the Newsbank database for a 10-year period covering 2009 to 2019. We selected the time frame based on developer timelines for both proposals as well as exploratory searches on Newsbank for articles with keywords on either end of the time frame. For the MBT case, we did three separate searches to account for frequent misspellings ("millennium," "millenium," and "milennium"). These three searches yielded a total of 758 articles and letters to the editor. For the GPT case, we conducted three searches as well since *The Bellingham Herald* produces a print and web edition of its paper, as well as maintaining an active news blog. Searches for the keyword phrase "gateway pacific" in these three platforms yielded 1561 articles, though many of these were duplicates since articles were often printed in both web and print versions of the paper.

We then undertook a thorough data cleaning process. First, duplicate articles were removed from the corpus. Then, we removed all punctuation, proper nouns, numbers, white space, and stop words (words that stand alone and do not have any meaning by themselves), as well as converting all

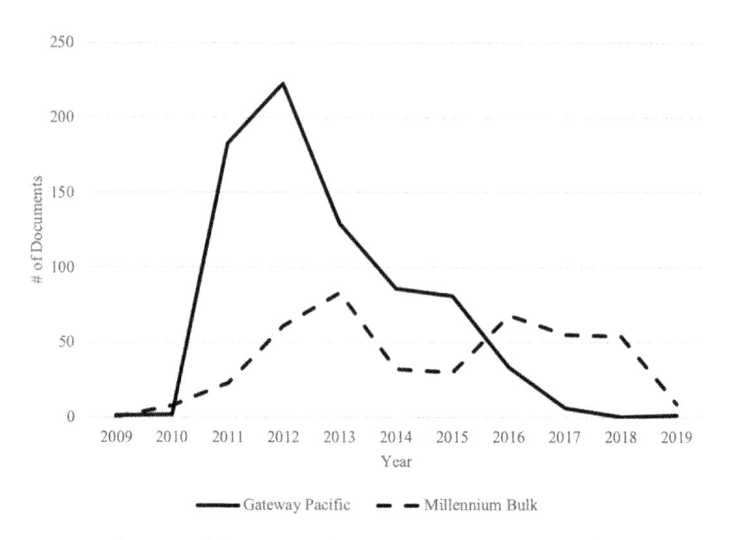

Figure 14.2 Distribution of documents by year and case number of newspaper articles by year for Gateway Pacific and Millennium Bulk, 2009–19.

text to lowercase and lemmatizing all words to reduce words to their common root (lemmatizing was conducting using the *NLTK* package). Bigrams (two-word phrases such as "coal train" and "climate change") were identified automatically then assessed and edited qualitatively to insure important and relevant ones were included in the data. The cleaning process yielded a dataset of 745 documents for the GPT case and 423 documents for the MBT case. The GPT data had a mean word count of 347.4 (standard deviation: 278.1), and the MBT data had a mean word count of 314.5 (standard deviation: 226.4). Fig. 14.2 shows the number of documents per year for each case for the final, cleaned sample.

Topic modeling using LDA

With two cleaned datasets, we then employed an unsupervised machine learning method of topic modeling known as Latent Dirichlet Allocation (LDA), using the Python package *gensim*. LDA identifies latent semantic themes in text data across a group of documents (Blei et al., 2003)—in our case, the cleaned newspaper articles and letters to the editors. In particular, each document is considered to have a distribution of latent topics, where each topic is recognized as a distribution of words that often occur together (Blei et al., 2003). LDA does not account for the order of words or their relative proximity to each other. This method is relatively new for environmental social scientists, but has been used increasingly in various social science disciplines (Bohr & Dunlap, 2018; Cody et al., 2017; Giordono et al., 2020).

LDA requires the researcher to select the number of topics for the LDA algorithm to discover. Though guidelines have been suggested by some as far as choosing the number of topics (Zhao et al., 2015), the researcher must be knowledgeable about the issue under study and be able to make an informed decision about how many topics are likely to exist. A choice of fewer topics will lead to results with more general themes, while a larger number of topics will generate more distinct themes. In our case, we assessed 15, 20, 25, 30, and 35 topic models and found that the 30-topic model, as compared with the other models, was rich and highly interpretable. We also examined the various models using quantitative metrics (such as topic coherence score), though these metrics were not as useful as our qualitative assessment. Last, we performed a "ground-truthing" procedure that involved reading a set of 50 randomly selected original documents for each case and making sure the topics assigned by the LDA procedure made sense.

Findings

The first two authors began by qualitatively evaluating the topics in the 30-topic model for each of the two datasets. Each topic in each model was first assessed for whether or not it was coherent; this was achieved through extensive study and discussion of the 30 most relevant words or phrases the algorithm associated with each topic. Then, the primary authors gave each topic a name.

Tables 14.1 and 14.2 present the named topics for the GPT and MBT models, respectively, alongside the top 10 most relevant words or phrases (bigrams) for each. The topic model for the GPT dataset had no topics that the researchers deemed unclear, while the topic model for the MBT dataset had two topics (#1 and #27) that the researchers felt were not coherent enough to name. Additionally, a handful of topics in both models seemed unimportant or irrelevant and were thus removed from the analysis. For example, in Table 14.1, topic 20 seems to describe logistics related to general news reporting, and topic 21 relates generally to the county council elections process. This process left 24 namable and relevant topics for both the GPT and MBT datasets. It's worth noting that the LDA algorithm does not assess sentiment—that is, it does not assess whether the

Table 14.1 Top 10 terms/phrases by topic for Gateway Pacific documents.

Topic	Topic name	Top 10 words/phrases by frequency within topic
1	Tribal rights	Lummi, tribe, nation, treaty, tribal, lummi_nation, protect, corp, army, land
2	Tax benefits	Tax, school, cost, pay, office, property, board, government, district, money
3	*SSA Marine*[a]	SSA, marine, company, ssa_marine, work, site, cole, spokesman, operation, firm
4	Environmental planning	Plan, land, resource, pier, acre, site, reserve, herring, build, management
5	Impact studies	Impact, study, health, potential, issue, concern, pacific, call, include, effect
6	Existing industry	Oil, ferndale, refinery, area, industry, north, include, place, bp, exist
7	Rail traffic	Train, rail, bnsf, railroad, traffic, track, capacity, build, increase, day
8	Union support	Job, local, labor, family, construction, industry, create, worker, industrial, union
9	Coal states commerce	Export, montana, mine, ton, accord, ship, gpt, march, company, percent
10	Energy markets and demand	Energy, report, power, demand, market, price, chinese, industry, company, peabody
11	Park impact	Side, road, park, close, cross, improvement, plan, work, access, transportation
12	Need for economic study	Study, estimate, impact, economic, result, facility, question, percent, business, add
13	Permitting process	Permit, decision, process, review, require, application, law, issue, complete, development
14	*Monday meeting*[a]	Member, decision, stand, today, monday, fight, fact, thing, victory, percent
15	Water and air pollution	Water, dust, air, pollution, clean, river, coal_dust, wind, level, environment
16	Environmental lawsuit	Attorney, file, group, june, lawsuit, pier, clear, stop, openurl_link, link
17	Exporting emissions	Export, china, northwest, world, pacific, ship, burn, carbon, country, fuel
18	Planning for future	Community, future, support, protect, plan, public, change, work, oppose, continue
19	Regional export traffic	Cargo, line, ship, train, resident, south, local, run, group, handle
20	*News logistics*[a]	Time, story, information, article, include, report, issue, statement, question, give
21	*Council campaign*[a]	Council, campaign, email, county_council, july, commission, interest, read, page, group

Table 14.1 Top 10 terms/phrases by topic for Gateway Pacific documents.—cont'd

Topic	Topic name	Top 10 words/phrases by frequency within topic
22	Waterfront redevelopment	City, council, mayor, waterfront, resolution, ordinance, member, cost, city_council, approve
23	Attracting business	Business, service, work, provide, issue, program, fee, time, development, experience
24	*Political elections*[a]	Candidate, district, election, vote, voter, issue, support, race, party, run
25	Army corps review	Corp, letter, request, fish, vessel, corps, impact, statement, marine, fishing
26	Citizen political power	Legal, local, government, initiative, people, system, law, bill, court, authority
27	Climate emissions	Letter, agency, climate, ecology, change, lead, governor, note, june, asia
28	Local economic benefit	Job, community, support, good, economy, gpt, live, opportunity, great, people
29	*Thursday meeting*[a]	People, meeting, speak, hear, concern, opponent, meet, event, mitigate, lot
30	EIS process	Comment, process, public, agency, eis, scoping, write, begin, draft, submit

[a]Unclear and unimportant/irrelevant topic names italicized.

Table 14.2 Top 10 terms/phrases by topic for Millennium Bulk documents.

Topic	Topic name	Top 10 words/phrases by frequency within topic
1	*Unclear topic*[a]	Area, concern, economic, lead, time, create, call, long, impact, question
2	Traffic impact	Industrial, area, traffic, intersection, transportation, congestion, include, access, road, growth
3	Brownfield cleanup	Clean, cleanup, alcoa, plan, property, alternative, aluminum, reynolds, cost, contaminate
4	Climate emissions	Burn, emission, global, carbon, Change, climate, plant, reduce, energy, greenhouse
5	Water pollution	Water, quality, impact, resource, federal, pollution, water_quality, rail, vessel, act
6	Legal challenges	Court, sublease, decision, file, argue, lawsuit, case, deny, judge, land
7	Seattle Council involvement	City, letter, council, seattle, government, issue, people, member, agency, political
8	Permitting process	Decision, law, board, deny, appeal, rule, review, shoreline, cowlitz_county, traffic
9	Local economic benefit	Tax, support, issue, pay, local, percent, process, change, long, school
10	Health impact study	Health, study, community, assessment, public, committee, recommendation, meet, potential, member

Continued

Table 14.2 Top 10 terms/phrases by topic for Millennium Bulk documents.—cont'd

Topic	Topic name	Top 10 words/phrases by frequency within topic
11	Local jobs	Work, longview_wa, put, people, benefit, link, openurl_link, employee, high, june
12	Agency delay	Ecology, agency, time, move, department_ecology, department, work, director, consultant, delay
13	Attracting business	Business, bring, create, industry, side, town, talk, plant, million, dollar
14	Diesel train emissions	Percent, increase, train, emission, risk, diesel, locomotive, cancer, bnsf, standard
15	Union support	Community, support, local, people, union, work, family, president, leader, great
16	EIS process	Study, comment, impact, draft, public, eis, statement, environmental_impact, release, final
17	Improve rail system	Rail, train, plan, improvement, railroad, corridor, expansion, line, traffic, transport
18	Public hearings	People, hearing, meeting, hear, speak, hold, opponent, center, anti, sign
19	Commissioner battle	Commissioner, official, land, property, acre, cowlitz_county, develop, accord, lease, add
20	*Political elections*[a]	District, vote, bill, candidate, house, campaign, republican, democrat, rep, elect
21	Regulatory jurisdiction	Review, official, northwest, agency, regulator, corp, federal, scope, group, conduct
22	*News logistics*[a]	Provide, news, give, process, write, daily, editorial, economic, opinion, daily_news
23	Coal states commerce	Wyoming, federal, administration, montana, nation, commerce, lighthouse, south, basin, foreign
24	Asia coal export	Export, energy, mine, china, market, accord, demand, percent, ton, report
25	Coal dust	Dust, train, coal_dust, car, day, air, area, ship, load, water
26	*First public hearing*[a]	Plan, large, tuesday, facility, export, daily, ton_coal, ton, application, begin
27	*Unclear topic*[a]	Time, good, back, open, problem, point, find, lot, long, thing
28	Dock expansion	Dock, river, columbia, construction, official, columbia_river, public, monday, include, riverkeeper
29	Energy markets and demand	Future, world, country, part, trade, agreement, power, continue, industry, clean
30	Regional export proposals	Base, oregon, group, proposal, ambre, pacific, call, st, daily_news, arch

[a] Unclear and unimportant/irrelevant topic names italicized.

topic relates most with support or opposition for the coal export proposal. Rather, the algorithm simply groups concomitant words together into topics. In some cases, as with the "union support" topic that appears in both cases (#8 in GPT and #15 in MBT), the sentiment is fairly straightforward. However, many topics appear fairly neutral—for example, topic 25 in Table 14.1 relates to coverage or discussion about the environmental review process.

Figs. 14.3 and 14.4 present a visualization of the relative prominence of each of the 24 nameable topics in the corpus for both the GPT and MBT datasets. For each topic in the figure, the proportion of each topic's words across all documents combined was calculated. Another way of describing this is that a proportion of 0.3 means that 30% of all documents contained that topic. Fig. 14.3 suggests that for the GPT proposal, the top five most prevalent topics were existing industry, environmental planning, water and air pollution, union support, and planning for the future. Fig. 14.4 suggests the top five most common topics related to the MBT proposal were regional export proposals, dock expansion, regulatory jurisdiction, Asia coal export, and union support.

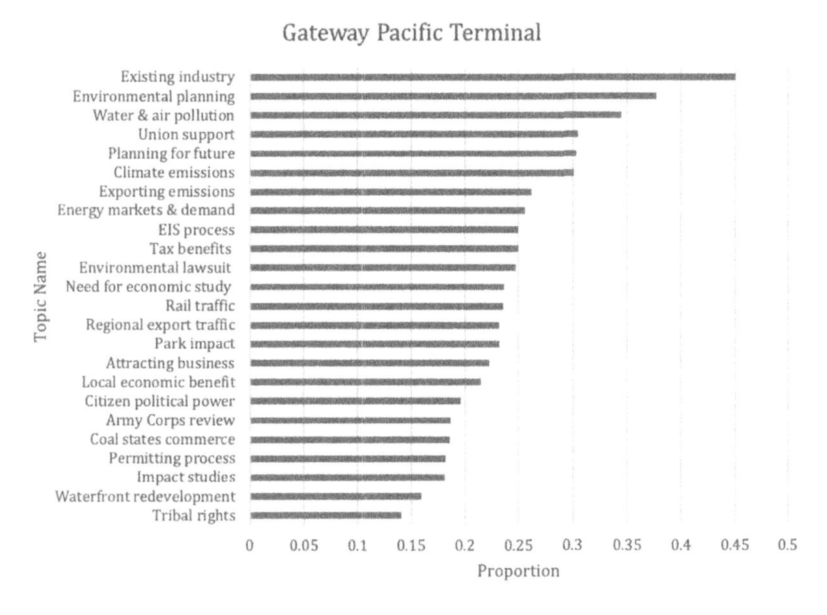

Figure 14.3 Proportion of topics across relevant documents for Gateway Pacific Terminals newspaper articles and letters to the editor (unclear topics and topics irrelevant to analysis excluded).

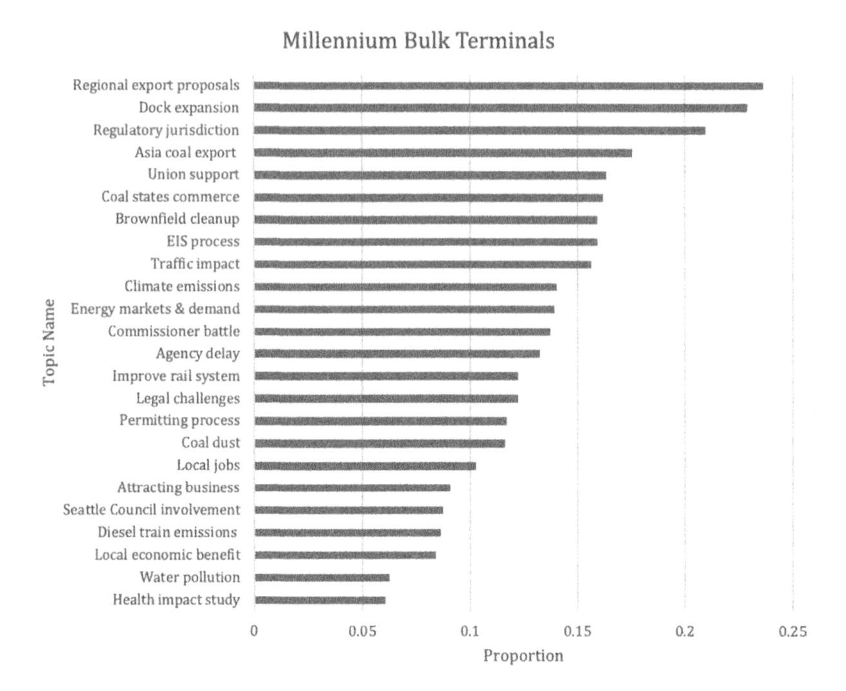

Figure 14.4 Proportion of topics across relevant documents for Millennium Bulk Terminals newspaper articles and letters to the editor (unclear topics and topics irrelevant to analysis excluded).

To provide a basis for comparison between the two cases, as well as to get a sense of balance between local versus scaled-up themes in the debate, we next sorted the 24 named topics into five categories: Environment & Health, Local Impacts & Planning, Economic Benefits, Review Processes, and Scaled-up Issues.

Table 14.3 provides the categorized topics for each case. We assigned topics to the Environment & Health category that had to do with potential impacts from the coal terminals to the local environment, natural resources, waterways, wetlands, and public health. We assigned topics to Local Impacts & Planning that had to do with potential impacts to local infrastructure, traffic, and planning processes. The Economic Benefits category is fairly self-explanatory, containing topics that related to different ways the proposals would benefit the local community economically. The largest category, Review Processes, contains topics that related to the various assessments and review processes conducted (or that were being called for) specifically about the proposals—these include permit review by various

Table 14.3 Categorized topics by case.

Topic category	Gateway Pacific	Millennium Bulk
Environment & Health	Water and air pollution	Water pollution Dock expansion Diesel train emissions Coal dust
Local Impacts & Planning	Waterfront redevelopment Planning for future Environmental planning Existing industry Rail traffic Park impact	Traffic impact Improve rail system Brownfield cleanup
Economic Benefits	Local economic benefit Attracting business Union support Tax benefits	Local economic benefit Attracting business Union support Local jobs
Review Processes	EIS process Impact studies Permitting process Army corps review Need for economic study Environmental lawsuit Citizen political power	EIS process Health impact study Permitting process Legal challenges Agency delay Regulatory jurisdiction Commissioner battle Seattle Council involvement
Scaled-up Issues	Energy markets and demand Tribal rights Coal states commerce Regional export traffic Climate emissions Exporting emissions	Energy markets and demand Asia coal export Coal states commerce Regional export proposals Climate emissions

agencies, the federal and state Environmental Impact Statements (EIS), health impact studies, related lawsuits and other legal challenges, engagement by outside organizations (such as the Seattle City Council), contention over regulatory jurisdiction, disagreements within agencies (as was the case between county commissioners in the MBT case), and perceived delays in the permitting processes. Last, we assigned topics to the Scaled-up Issues category if they related to issues that were clearly broader than local, such as topics about global energy markets, interstate commerce, climate emissions, and so on. Below, we structure our discussion of findings by our research questions.

How do the two cases differ in terms of the themes that were most prominent?

First and foremost, we note differences in the Environmental & Health and Local Impacts & Planning categories between the two cases. While keywords related to water pollution appeared in both cases, the topic model for MBT indicated additional attention in three themes: dock expansion, diesel train emissions, and coal dust. The MBT proposal included a proposal to build a second ship loading dock in addition to the existing dock that was used by the previous Alcoa aluminum facility, and there was great concern about the impact of building this second dock (including required dredging) on the Columbia River and its aquatic species. There was also a heightened dialogue in the MBT case about the impacts of coal dust from trains and the railyard, including how dust coming off moving trains (which would travel a great distance along the Columbia River) might affect both human and aquatic species. Similarly, concern over diesel exhaust and fuel-based particulates such as PM2.5 (Galvis et al., 2013) featured more prominently in the debate over the MBT proposal, possibly spurred by a 2013 legal brief that recommended states regulate the particulate matter emanating from Powder River Basin coal trains (Trimming, 2013) and a 2015 study conducted along the Columbia River railway concluded that open-top coal trains produced almost double the amount of PM2.5 than freight trains (Jaffe et al., 2015). That is, while both terminals intended to export coal, the 2-year lag between the proposals meant that the timing of the above referenced legal brief and study could have had more of an impact on MBT debates than on GPT deliberations.

The two cases differed significantly in terms of how many topics/themes arose in the Local Impacts & Planning category, with far more arising in the GPT case. There were three themes in this category for MBT, related to concerns about impacts of increased numbers of train on local traffic patterns, emphasis on how the project could or should improve the rail system, and the feeling that the project would help clean up an existing, defunct industrial site. Conversely, in the GPT case, we found a much greater emphasis on themes relating to local planning, including how the project could impact the waterfront redevelopment project in the city of Bellingham, the waterfront parks, and the need to consider the project in the cities' larger vision for the future. Concerns about train traffic emerged in the GPT case, but were more geared toward general concern over the number of trains that would be passing through the waterfront each day, rather than about direct impacts on local vehicle traffic patterns. The concept that GPT would add to existing industry in the area (including an oil refinery, shipping docks, and an aluminum plant) had both positive and negative connotations.

Themes of economic benefit appeared in both GPT and MBT topic models, with three of the four themes seemingly the same (local economic benefit, attracting business, and union support). The two cases each had one unique theme in this category—tax benefits for GPT and an emphasis on local jobs in the MBT case.

The largest category for both cases was that pertaining to topics about the review processes, including the EIS processes, various impact studies, the permitting process, and legal challenges. The GPT case included a theme wherein an economic study was called for by citizens (presumably to counteract the narrative of the developer that it would bring economic benefit to the area), as well as a theme relating to the power of citizens to influence the process through voting and attending meetings. Conversely, the MBT case highlighted discontent about how long the review process took (agency delay), contention over which agency should have jurisdiction over the decision process, disagreement between local county commissioners, and involvement from outside organizations.

How did local versus "scaled-up" concerns feature in the GPT and MBT debates?

We now examine the last category of themes that emerged in our topic models and which represent "scaled-up" themes of contention. Both GPT and MBT emphasized energy markets and demand, though the theme of coal export to Asia specifically arose more prominently in the MBT case. Debate over about interstate commerce and the right of interior coal states to export their products through coastal states emerged in both GPT and MBT; however, it's worth noting that this issue developed into a lawsuit with several coal states bringing a case against Washington's governor and Department of Ecology after a key permit was denied to the MBT developer. Last, conflict over the contributions of both export projects to global climate emissions emerged, with an additional theme related to concern about "exporting emissions" emerging in the GPT case.

Some key differences between the two cases in terms of these "scaled-up" issues include the theme of tribal rights in the GPT case and differently emphasized discourse over regional export between the two cases. The issue of tribal rights featured relatively prominently in GPT, with the local Lummi Nation asserting rights to traditional fishing grounds and successfully arguing the project would hamper those rights due to its proposed location and its potential impact on several species. While this topic pertains to a local tribal nation's contention, it also included keywords that suggested broader contention over tribal sovereignty, fishing rights, and other impacts. A second difference is that while GPT discourse focused on concern over increased export traffic in the broader region, including increased shipping traffic and rail traffic, MBT discourse focused more on the various proposals for new export facilities around the region, with keywords including place names and project names for previous proposals. This suggests that MBT discourse was more connected than GPT to broader awareness of increasing interest in the Pacific Northwest as a potential export hub.

Overall, in terms of our research question about how local versus broader concerns featured in the GPT and MBT debates, we find that both local and "scaled-up" themes were present in both cases, but perhaps more so in the MBT case.

Conclusion: implications for fossil fuel export in the Pacific Northwest

We used a comparative case study design with computer-assisted topic modeling to better understand the community response to coal export terminals in Washington State. It might be easy to suggest that coal export projects failed because of the pushback from local and regional environmental groups. However, our analysis uncovered a much more nuanced picture. We noted that in the GPT case, community concerns about the impact on the local waterfront development and damage to the local environment and fisheries came to the fore, while in the MBT case, greenhouse gas emissions and issues surrounding exporting fossil fuels to Asia rose to the top. In both cases, concerns about who had jurisdiction became a focus—the Lummi Nation asserted treaty rights in the GPT case and, in the MBT case, representatives of six interior states including Montana and Wyoming sued Governor Jay Inslee and the State of Washington to safeguard their right to ship coal abroad. Classifying the result of either terminal project as the result of NIMBY or even NIABY thinking would certainly obfuscate important details about the differences between the responses to the projects.

What can be said about the broader regional response to fossil fuel export proposals? We do know that these two terminals would have affected wide-ranging communities, and those communities made their voices heard. In the case of GPT, more than 124,000 comments were received during scoping process ("More than 124,000 Comments Received on Scope of Environmental Review for Proposed Cherry Point Export Terminal," 2013). At Longview, over 3000 people showed up at just one MTB hearing. The Seattle City Council weighed in on MBT, voting unanimously to oppose the terminal project. The results of our topic models suggest that while local concerns were numerous in both cases, contestation over both GPT and MBT included several "scaled-up" themes, many of which were similar between the two cases. Particularly, concern (but also debate) about the contribution of US coal burned abroad to global greenhouse gas emissions was a major focus, as was contestation over the rights of coal states and awareness of the implications of new regional coal terminals on global energy markets. While our analysis cannot assess the processes underlying social movement "scale shift" (Boudet, 2011; Tarrow & McAdam, 2004), we feel these results suggest not only local awareness of larger-than-local dimensions to the coal terminal debate but also a high likelihood that environmental groups, political leaders, and other opponents have succeeded in proliferating narratives against fossil fuel exports in this region by invoking broader themes that make the issue salient to individuals outside the host communities.

Our study bears some limitations. First, our data consist of articles from just one local newspaper in each location. We do not explore articles or letters to the editor in any regional papers like the Seattle Times or the Portland Tribune; doing so would represent yet another step in assessing scaling of the movement. Such an analysis might also help us understand the diffusion of information across the state and potential coordination of activities between groups both supporting and opposing the terminal projects. Second, we do not examine newspaper articles as compared with letters to the editor and opinion editorials—due to the limitations of LDA and the lack of metadata, these document types are assessed together. Last, we do not assess how the relative prominence of different topics generated by the LDA algorithm fluctuated over time. Future research that explores the timing of events related to the terminal projects and hearings, and the publication of articles in news media across the region, would help us better understand the role the media played in community response to proposed export terminals.

Our research highlights one application of computer-assisted topic modeling for social scientists studying community response to energy infrastructure projects. While Grimmer and Stewart warn that LDA topic modeling has both "promise and pitfalls" (Grimmer & Stewart, 2013), we propose this method has great potential for assessing large collections of texts relatively quickly. For researchers studying community and public responses to energy infrastructure, a range of document types could be assessed, including newspaper articles and letters to the editor, public testimony and comments from environmental review processes, legal and governmental agency documents, and even possibly online content or press releases from organizations. Ultimately, LDA and other topic modeling algorithms open new opportunities for energy social science, and could contribute substantially to the state of knowledge on social dimensions of fossil fuels and fossil fuel export worldwide.

References

Allen, M., Bird, S., Breslow, S., & Dolsak, N. (2017). Stronger together: Strategies to protect local sovereignty, ecosystems, and place-based communities from the global fossil fuel trade. *Marine Policy, 80*, 168–176. http://www.sciencedirect.com/science/article/pii/S0308597X16306704.

Barrera, P. (2021). *Coal outlook 2021*. Investing News Network. https://investingnews.com/daily/resource-investing/industrial-metals-investing/coal-investing/coal-outlook/.

Bartlett, L., & Vavrus, F. (2017). Comparative case studies: An innovative approach. *Nordic Journal of Comparative and International Education (NJCIE), 1*(1). https://doi.org/10.7577/njcie.1929

Benford, R. D., & Snow, D. A. (2000). Framing processes and social movements: An overview and assessment. *Annual Review of Sociology, 26*, 611–639. https://doi.org/10.1146/annurev.soc.26.1.611

Blei, D. M., Ng, A. Y., & Jordan, M. I. (2003). Latent dirichlet allocation. *Journal of Machine Learning Research, 3*(4–5), 993–1022.

Bohon, S. A., & Humphrey, C. R. (2000). Courting LULUs: Characteristics of suitor and objector communities. *Rural Sociology, 65*(3), 376–395. https://doi.org/10.1111/j.1549-0831.2000.tb00035.x

Bohr, J., & Dunlap, R. E. (2018). Key topics in environmental sociology, 1990–2014: Results from a computational text analysis. *Environmental Sociology, 4*(2), 181–195. https://doi.org/10.1080/23251042.2017.1393863

Boudet, H. S. (2011). From NIMBY to NIABY: Regional mobilization against liquefied natural gas in the United States. *Environmental Politics, 20*(6), 786–806. https://doi.org/10.1080/09644016.2011.617166

Brownell, S., Miller, K., O'Guin, S., Reider, M., & Ward, S. (2012). *Cherry Point coal trains: Environmental impact assessment.*

City of Vancouver. (2016). *Staff report No. 141A-16.*

Cody, E. M., Stephens, J. C., Bagrow, J. P., Dodds, P. S., & Danforth, C. M. (2017). Transitions in climate and energy discourse between Hurricanes Katrina and Sandy. *Journal of Environmental and Social Sciences,* 7(1), 87–101. https://doi.org/10.1007/s13412-016-0391-8

Dear, M. (1992). Understanding and overcoming the NIMBY syndrome. *Journal of the American Planning Association, 58*(3), 288–300. https://doi.org/10.1080/01944369208975808

Energy Information Administration. (2017). *Future coal production depends on resources and technology, not just policy choices—today in energy.* https://www.eia.gov/todayinenergy/detail.php?id=31792.

Energy Information Administration. (2018). *Washington - state energy profile analysis.* https://www.eia.gov/state/analysis.php?sid=WA#115.

Energy Information Administration. (2019). *U.S. Number and capacity of petroleum refineries.* https://www.eia.gov/dnav/pet/pet_pnp_cap1_dcu_nus_a.htm.

Energy Information Administration. (2021). *Quarterly coal report.* https://www.eia.gov/coal/production/quarterly/index.php.

Erickson, P., Lazarus, M., & Piggot, G. (2018). Limiting fossil fuel production as the next big step in climate policy. *Nature Climate Change, 8*(12), 1037–1043. https://doi.org/10.1038/s41558-018-0337-0

Fox, T. (2018). *Opinion | Washington state should stop blocking planned coal export terminal.* The New York Times.

Galvis, B., Bergin, M., & Russell, A. (2013). Fuel-based fine particulate and black carbon emission factors from a railyard area in Atlanta. *Journal of the Air and Waste Management Association, 63*(6), 648–658. https://doi.org/10.1080/10962247.2013.776507

Gerring, J. (2004). What is a case study and what is it good for? *American Political Science Review, 98*(2), 341–354. https://doi.org/10.1017/S0003055404001182

Giordono, L., Boudet, H., & Gard-Murray, A. (2020). Local adaptation policy responses to extreme weather events. *Policy Sciences, 53*(4), 609–636. https://doi.org/10.1007/s11077-020-09401-3

Grimmer, J., & Stewart, B. M. (2013). Text as data: The promise and pitfalls of automatic content analysis methods for political texts. *Political Analysis, 21*(3), 267–297. https://doi.org/10.1093/pan/mps028

Hasegawa, K. (2010). *A comparative study of social movements for a post-nuclear energy era in Japan and the USA.* Springer Science and Business Media LLC. https://doi.org/10.1007/978-0-387-09626-1_3

Hellegers, D. (2021b). *The thin green line is people.* Washington State University. https://labs.wsu.edu/thethingreenlineispeople/.

Hellegers, D. (2021a). *City council resolutions. The thin green line is people.* Washington State University. https://labs.wsu.edu/thethingreenlineispeople/city-council-resolutions/.

High court upholds DNR denial of sublease for Longview coal dock. (2020). The Daily News. https://tdn.com/news/local/high-court-upholds-dnr-denial-of-sublease-for-longview-coal-dock/article_f4f37585-19ae-5dee-913e-22a79a400cfc.html.

Jaffe, D., Putz, J., Hof, G., Hof, G., Hee, J., Lommers-Johnson, D. A., Gabela, F., Fry, J. L., Ayres, B., Kelp, M., & Minsk, M. (2015). Diesel particulate matter and coal dust from trains in the Columbia River Gorge, Washington State, USA. *Atmospheric Pollution Research, 6*(6), 946–952. https://doi.org/10.1016/j.apr.2015.04.004

Joppke, C. (1991). Social movements during cycles of issue attention: The decline of the anti-nuclear energy movements in West Germany and the USA. *British Journal of Sociology, 43*. https://doi.org/10.2307/590834

Koopmans, R., & Olzak, S. (2004). Discursive opportunities and the evolution of right-wing violence in Germany. *American Journal of Sociology, 110*(Issue 1), 198–230. https://doi.org/10.1086/386271

LINGO. (2019). *Leave fossil fuels in the ground*. http://leave-it-in-the-ground.org/.

McClure, R. (2021). *InvestigateWest: Activists thwart fossil fuel projects*. OBP. https://www.opb.org/article/2021/01/18/decarbonizing-cascadia-fossil-fuels/?utm_medium=email&utm_term=Read%20the%20story&utm_content=First%20Look%20Jan%2019%202021%20CID_a0fdecf8832c6a3dcec85d7225d016d9&utm_source=firstlook&utm_campaign=First%20Look%20Jan%2019%202021.

More than 124,000 comments received on scope of environmental review for proposed Cherry Point export terminal. (2013). San Juan Islander. https://sanjuanislander.com/news-articles/environment-science-whales/1837/more-than-124-000-comments-received-on-scope-of-environmental-review-for-proposed-cherry-point-export-terminal.

Piggot, G. (2018). The influence of social movements on policies that constrain fossil fuel supply. *Climate Policy, 18*(7), 942–954. https://doi.org/10.1080/14693062.2017.1394255

Popper, F. J. (2010). The environmentalist and the LULU. *Environment: Science and Policy for Sustainable Development, 27*(2), 7–40. https://doi.org/10.1080/00139157.1985.9933448

Rice, N. (2011). *Northwest coal port ignites controversy*. High Country News. https://www.hcn.org/blogs/goat/northwest-coal-port-ignites-controversy.

Richards, M. (2017). Labor urges passage of Millennium Bulk Terminals permit to boost economy, create jobs. *The Lens*. https://thelens.news/2017/10/12/labor-urges-passage-of-millennium-bulk-terminals-permit-to-boost-economy-create-jobs/.

Schultz, S. (2011). *Seeking a Pacific Northwest gateway for U.S. Coal*. National Geographic News. https://www.nationalgeographic.com/science/article/111020-coal-port-pacific-northwest.

Schwartz, R. (2015). *Lummi nation asks Army Corps to reject Cherry Point coal terminal*. The Bellingham Herald.

Seattle Times Staff. (2017). *EPA bashes early environmental study of Longview coal terminal*. The Seattle Times. https://www.seattletimes.com/seattle-news/environment/epa-bashes-early-environmental-study-of-longview-coal-terminal/.

Sightline Institute. (2018). *The thin green line*. https://www.sightline.org/research/thin-green-line/.

Snow, D. A., & Benford, R. D. (1992). Master frames and cycles of protest. In *Frontiers in social movement theory*.

Tarrow, S., & McAdam, D. (2004). Transnational protest and global activism. In *In transnational protest and global activism*. Rowman & Littlefield Publishers.

Trimming, T. R. (2013). *Derailing powder river basin coal exports: Legal mechanisms to regulate fugitive coal dust from rail transportation* (Vol. 6, pp. 321–345).

U.S. Army Corps of Engineers. (2016). Millennium bulk terminals environmental impact Statement. In *US Army Corps of Engineers*.

U.S. Census Bureau. (2020). *QuickFacts: Whatcom county, Washington; Bellingham city, Washington*. https://www.census.gov/quickfacts/fact/table/whatcomcountywashington,bellinghamcitywashington,WA/PST045219.

Useem, B., & Zald, M. N. (1982). From pressure group to social movement: Organizational dilemmas of the effort to promote nuclear power. *Social Problems*, 144–156. https://doi.org/10.2307/800514

Washington election results 2020: Live results by county. (2021). NBC News. https://www.nbcnews.com/politics/2020-elections/washington-results.

Whatcom County. (2011). *Emails pertaining to gateway pacific project for december 17-23, 2011.* https://www.whatcomcounty.us/DocumentCenter/View/3075/Comprehensive-List—December-17—December-23-2011-PDF.

Williams-Derry, C., & de Place, E. (2017). *Northwest coal exports: The end is nigh.* https://www.sightline.org/2017/01/05/northwest-coal-exports-the-end-is-nigh/.

Wysham, D. (2016). *This city just banned virtually all new dirty-energy infrastructure.* The Nation. https://www.thenation.com/article/archive/this-city-just-banned-virtually-all-new-dirty-energy-infrastructure/.

Yin, R. K. (2017). *Case study research and applications: Design and methods.* Sage Publications.

Zhao, W., Chen, J. J., Perkins, R., Liu, Z., Ge, W., Ding, Y., & Zou, W. (2015). A heuristic approach to determine an appropriate number of topics in topic modeling. *BMC Bioinformatics, 16*(13). https://doi.org/10.1186/1471-2105-16-S13-S8

PART V

The future of fossil fuel export in an era of energy transition

CHAPTER 15

Social dimensions of fossil fuel export: summary of learnings and implications for research and practice

Shawn Hazboun[1] and Hilary Boudet[2]
[1]Graduate Program on the Environment, The Evergreen State College, Olympia, WA, United States;
[2]Sociology, School of Public Policy, Oregon State University, Corvallis, OR, United States

This edited volume highlights research on public response to fossil fuel export projects and required transportation infrastructure around the world. While energy social science is robust (Sovacool, 2014; Sovacool et al., 2018), the literature to date has largely focused on sites of energy extraction, leaving a gap in knowledge regarding public response to energy export. Through this volume, we make the case that the movement of fossil fuels deserves more scholarly attention. The chapters in this volume suggest that public response to fossil fuel export is similar in many ways to energy extraction in terms of its impacts to host communities and the themes that arise in public debate. Yet, there are features of fossil fuel export that make it somewhat unique. In this concluding chapter, we summarize learnings and consider implications for future research and practice.

Climate is an increasing concern in energy export debates

The research in this volume suggests that public response to fossil fuel export projects (as well as other types of fossil fuel projects) must be contextualized within growing awareness of how energy production contributes to climate change, the threat of which now looms larger than ever before (Masson-Delmotte et al., 2021). As Piggot and Erickson discuss in Chapter 3, nations are beginning to consider various "supply side" policies to limit the flow of fossil fuels, and some are choosing to stop investing in any new fossil fuel infrastructure. An example of this is highlighted in Chapter 6, in which Widener describes New Zealand's 2017 ban on new offshore oil exploration, instituted as part of the country's climate mitigation effort. Yet, how are impacted communities actually responding?

Public Responses to Fossil Fuel Export
ISBN 978-0-12-824046-5
https://doi.org/10.1016/B978-0-12-824046-5.00015-1

The chapters in this volume suggest that localized concerns remain at the forefront of energy siting debates. However, they also highlight how global climate change is increasingly a concern both within and beyond host communities. Proposals for export-specific infrastructure seem particularly tenuous, perhaps because the global implications are more easily intuited. Communities facing an export proposal must not only weigh the impacts and benefits of the project to the local economy and environment but must also reckon with their own role in the unfolding climate crisis as they envision ship after ship of coal or liquefied natural gas leaving their harbor for overseas ports. As Hazboun and Boudet describe in Chapter 8, some coastal regions are strongly pushing back against fossil fuel export proposals, positioning themselves as guardians against a Pandora's Box of greenhouse gas emissions (Erickson et al., 2018; LINGO, 2021; Piggot, 2018; Sightline Institute, 2018). And climate change features prominently in some export siting debates, such as with the coal export terminals discussed in Chapter 14, wherein stakeholders as well as the formal environmental review considered direct (burning USA coal abroad) and indirect (transportation-related) greenhouse gas contributions from the proposed projects.

Yet, many communities and nations depend on fossil fuel export for jobs and economic development, and this can make it challenging to reject proposals for new or expanded export and transport facilities, even when climate action is strongly desired. Communities can become "overadapted" to extractive industries such that pivoting away seems nearly impossible (Freudenburg, 1992). In Chapter 7, Andersen et al. describe the case of Norway, where (unlike in the United States) the government and general public accepts the reality of climate change and also where jobs and profits from fossil fuel exports greatly benefit citizens. Norway is an example where fossil fuel activities are not highly controversial, and this is also partly because export activities mostly occur offshore where they are out of sight and mind for most citizens. By contrast, most of the case study chapters highlight proposals that caused prominent public debates, such as over the Canadian pipeline expansion project described in Chapter 10, the Canadian and Australian LNG terminals highlighted in Chapters 11 and 12, and the US coal terminals highlighted in Chapter 14.

Concern about climate change will likely play an increasingly large role in debates over fossil fuel export, alongside local concerns. The extent to which climate mitigation drives the debate, however, will vary based on region and country, how visible the export activities are, what fuel type is being exported, and how committed governments are to including fossil fuel supply in their climate policies.

Export routes present many opportunities for opposition

Fossil fuel export requires linearly extensive transportation components. These include thousands of miles of pipelines, railroads, and shipping lines—and, these transportation components are often more controversial than the actual shipping terminal. Energy companies are fairly dependent on *existing* routes and infrastructure—though new facilities can be built, they are often cost prohibitive, can be highly controversial, and can be dependent on political regimes, such as in the case of the high profile Keystone XL Pipeline in the United States. Keystone XL was proposed in 2008 to transport Canadian crude bitumen to refineries in the Gulf Coast region, experienced explosive opposition and was vetoed by President Obama in 2015, repermitted by President Trump, then finally canceled by the developer after President Biden cast it down in January 2021.

Unlike a refinery, mine, or energy plant, export routes cross multiple jurisdictions (nations, even) and pass through many communities, introducing many opportunities for opposition. Though geographic extensiveness is not unique to export—oil and gas plays also cross jurisdictions and impact many communities—one key difference is that if one community along an export route is successful in obstructing a component in the export route, then the entire route connecting commodity to market is jeopardized. Some communities are starting to take a stand against fossil fuel transportation—for example, several municipalities in the northwestern United States, including the major city of Portland, Oregon, have adopted ordinances that prohibit the expansion of fossil fuel transportation and storage infrastructure (Profita, 2019).

Several chapters in this volume focus on community response to the transportation infrastructure necessary to export fossil fuels, including oil trains (Chapter 13), pipelines (Chapters 9 and 10), and shipping (Chapter 8). These chapters highlight how support or opposition can vary based on cultural context, familiarity and experience with the industry, risk perceptions, and political standing, just as with other types of energy projects. In cases where the export route crosses indigenous land, the impacts may include infringements on nationhood, sovereignty, and treaty rights (such as in Chapter 10).

Communities along transportation routes generally do not receive the same type of benefits as communities hosting an extractive industry or export terminal, yet they are vulnerable to transportation-related risks. Pipeline leaks impact the environment and public health, and can even result in deadly explosions, such as the 2010 leaking pipeline explosion in San Bruno, California, which killed 8 people and injured over 50. According to ProPublica investigations, since 1986, "pipeline accidents have killed more than 500 people, injured over 4000, and cost nearly $7 billion in property damages" (Groeger, 2012). Pipelines may also cross the treaty lands of indigenous nations, introducing environmental risks and often challenging tribal sovereignty. Ship and rail transport also present significant risks to communities. There are countless examples around the world of oil tanker spills harming coastal economies and ecologies. Railway communities bear the risk of oil train derailment, like in the small town of Lac-Megantic, Quebec, where in 2013 a 73-car crude oil train derailed and exploded in the downtown area, killing 47 people. Coal and oil trains impact local air quality due to the diesel particulates they produce, and coal dust that escapes uncovered train cars poses a risk to waterways and air along the railway (Jaffe et al., 2015). Whatever the mode, it seems clear that transporting fossil fuels bears risks for those living along the route.

Furthermore, exporting fossil fuels typically requires access to a coast. This may make trade routes even more vulnerable to opposition because many people have special place attachments to coastal areas, including those who live by the coast (Perry et al., 2014) and those who only visit the coast (Tonge et al., 2015).

All of the above suggests that export is a particularly weak point for the energy industry as a whole and is vulnerable to many touchpoints of opposition, perhaps more so than for extraction.

Changing attitudes about natural gas likely to impact export

Though hydraulic fracturing has been controversial across producing nations, the export of natural gas seems to have comparatively broad support. In Chapter 4, Zanocco et al. find that support for LNG export increased in the United States from 2013 to 2017, and in Chapter 8, Hazboun and Boudet find more public support for exporting natural gas than for exporting oil or coal. Yet, this might change as public understanding of natural gas as a serious greenhouse gas increases. For decades, the natural gas industry successfully cultivated the idea that natural gas is a less environmentally harmful fossil fuel, and even environmental groups at one point heralded it as an essential "bridge fuel" (Delborne et al., 2020; Hazboun et al., 2021; Leber, 2021). But this is changing, and some governments have rejected new natural gas infrastructure proposals on these grounds—one example is the recent decision by the provincial government of Quebec, Canada, to reject a $14 billion proposal for an LNG terminal to export natural gas, citing impacts to the global climate as well as how investment in the terminal would prolong the energy transition (Fawcett, 2021; Woodside, 2021). Yet, other governments, even those considered environmentally progressive, continue to support new natural gas projects—for example, the province of British Columbia, Canada, has referred to LNG as a "win-win" for the economy and the climate (Chapter 11).

Attitudes about LNG export specifically could shift if social movement groups become more successful at framing it as a risky industry. LNG technology requires special facilities to cool and compress the gas, store it, and ship it. Communities hosting such facilities are subject to certain risks, including the potential for major fire and explosions (Baalisampang et al., 2019; Englund, 2021). LNG ships are sometimes referred to as "floating bombs," though the LNG industry has a fairly impressive safety record (Mokhatab et al., 2014). Interestingly, in the LNG case study chapters in this volume (Chapters 9, 11, 12), safety concerns were not a factor in community responses.

Some familiar patterns persist

The research in this volume echoes several findings of the energy impacts and public perceptions literature to date. First, the case study chapters highlight the ongoing importance of public participation in siting efforts, and also how a lack of meaningful participation and community engagement can lead to significant distrust and upheaval. And, as Loginova highlights in Chapter 9, communities even in far away and remote places are learning of international best practices for community engagement, yet feel excluded when these standards are not utilized by energy developers. Even when developers do have formal participation processes, they may simultaneously use strategies of exclusion and "non-participation," as Loginova finds. Rules about consulting with indigenous peoples when their land is impacted by proposed energy infrastructure—as discussed in Chapters 9 and 10—also do not guarantee adequate engagement or consent. Overall, these chapters underscore an ongoing need for better community inclusion in export and transport facility siting decisions.

Second, host community impacts for new export terminals can follow similar boomtown patterns of social disruption. In Chapter 12, Benham details the huge influx of workers (upwards of 15,000) as LNG plants were constructed in the Australian communities of Gladstone and Darwin. As is often the case with extraction booms, the construction-phase workforce is most often male-dominated and temporary, and this often raises crime and substance abuse rates in the community. However, more research is needed to understand how long these impacts last—once the construction phase is over, do export communities become "overadapted" to the extractive industry? Do they experience boom and bust patterns, as extraction communities do? The case studies in this volume do suggest that other disruptions would be ongoing even after construction, such as impacts to traffic patterns from increased rail activity.

Third, communities grapple with many similar concerns for export proposals as they do for extraction. Residents may be alarmed that a new export terminal might harm the local environment, create public health risks, and impact their sense of place. Yet, other residents may feel that an export facility would bring jobs, generate tax revenue, and boost the local and regional economy. Indeed, the familiar "environment versus jobs" dichotomy shows up throughout the cases in this edited volume. Additionally, concern about local governance processes and democratic accountability emerged in the case studies, echoing findings of the literature on public response to hydraulic fracturing (Bomberg, 2017).

The public opinion research included in this volume also suggests that partisanship and politics play a significant role in public perception of fossil fuel export, as it does with public opinion on fracking and other forms of extraction. However, natural gas export, as mentioned, will likely continue to receive support on both sides of the political spectrum for a while longer owing to its reputation as a cleaner fuel or a "bridge fuel."

Implications (and a few limitations) for research and practice

The research in this volume bears some implications for the growing field of energy social science, particularly research that is focused on public perceptions of energy development, siting controversies and community impacts. The first implication is simply that the transportation and export phases of fossil fuel production deserve more scholarly attention. There is a paucity of research on public responses to fossil fuel export, and while this volume attempts to fill in some of the gap, it is also a call for more research on this topic. Given the increasing focus on limiting the global supply of fossil fuels, community and public opposition to export (and import, for that matter) activities could escalate to the point of disrupting the global trade of coal, oil, and gas—indeed, this is happening along the west coast of the United States, where coal companies have failed at securing new export terminals to ship their product to market.

On that note, this volume indicates that the extraction and export of fossil fuels will increasingly be seen as a climate policy problem in years to come. Whereas governments have focused on reducing domestic consumption of fossil fuels, some are now starting to target supply-side activities (as with New Zealand's ban on offshore drilling) and others are halting new investments in fossil fuel infrastructure, thus reducing the problem of "carbon lock-in." As Guyliev describes in Chapter 2, the global process of decarbonization will be uneven across nations, and will likely be slower going in energy-rich countries like Russia, Australia, Canada, and the United States. Going forward, researchers studying community-level responses to new energy projects should analyze the extent to which supply-side concepts are part of local-level dialogue and discussion. Survey researchers, too, would do well to consistently include climate change attitudes in data collected about levels of support for different energy sources and energy-related activities—it has not been common practice to include climate change views in studies on public energy preferences (Clarke et al., 2019; Hazboun et al., 2020; Pierce et al., 2018). Practitioners, too, could gauge the climate attitudes of a proposed host community and should be prepared to address questions about the climate impact of a new energy project.

Another implication we draw from this book is for social scientists focusing on natural gas–related projects. Public views about natural gas seem to be complex, varying depending on what part of the natural gas production cycle is at hand. Chapters 4 and 8 as well as other polls suggest there is relatively high support for natural gas as an energy source and for exporting gas (Hazboun et al., 2020, 2021), yet there is controversy over hydraulic fracturing as well as emerging understanding about methane leaks during extraction and transport (Pandey et al., 2019; Weller et al., 2020).

Thus, researchers must be attentive to these nuances in public perception, the ways attitudes about natural gas are shifting, and the forces causing these shifts.

In addition to addressing an overall gap in the literature on public response to fossil fuel export, we also must point out that the literature that does exist is mainly from locations in western, educated, industrial, wealthy, and democratic countries. That is admittedly a limitation of this volume (though we tried, we were not as successful as we hoped in securing research from outside these locations) as well as the field of energy social science as a whole (Sovacool, 2014). Going forward, conducting and elevating research from other exporting—and importing—countries especially in Latin America, Asia, the Middle East, and Africa would help the field gain a more complete understanding of how the global fossil fuel trade impacts communities, and how different publics around the world are responding.

Additionally, future research needs to develop a more systematic way of studying the social impacts of different types of energy projects, whether extraction, transportation, export, or combustion for electricity. The majority of energy social science research, starting with the early studies on boomtowns in the 1970s and 1980s, has been reactive in nature—that is, projects get proposed or built, a debate ensues or an opposition movement begins, and then academics arrive on the scene to analyze the controversy. In this way, much of energy social science has been crisis-oriented (Rosa et al., 1988). The case studies in this volume also follow this pattern. The field needs to develop a more consistent and robust analysis of community response to energy projects across different contexts. Indeed, the field as a whole has a bias toward selecting study sites where there *is* controversy and opposition—a more systematic mode would entail studying a set of communities hosting the same type of infrastructure (an LNG terminal, for example) and examining the conditions or factors that do, or don't, provoke opposition. As an example of an alternative, more systematic way of studying community response, Giordono and colleagues' 2018 study of 53 community responses to wind energy siting in the Western US found that communities overall had relatively benign responses as compared with the level of opposition represented in the literature (Giordono et al., 2018).

This volume reflects two other limitations apparent in the research on public/community response to energy: (1) lack of coherent theory, and (2) methodological disconnect. As discussed in the introductory chapter, social scientists studying both general public opinion as well as community response have used a variety of theoretical orientations and frameworks over the decades, from the boomtown model and locally unwanted land uses, to environmental justice and risk perceptions, to social representations and framing, to resource mobilization and political opportunity structure. Each theory provides a contribution, yet the field has not come closer to developing a coherent theory of public response. Second, researchers tend to use one of two very different methodological approaches to study public response: survey research (reflected in Part 2 of this volume) and case study research (reflected in Part 3). While they can complement each other, it will take more intentional coordination and connection to draw big picture findings from across these two approaches. Examples of mixed-mode research do exist that highlight how survey and case study research, among other methods, can work well together (Hazboun, Howe, Layne Coppock, & Givens, 2020; Weible, Heikkila, & Carter, 2017), but these studies are usually the exception.

Closing words

Our energy system stands at a crossroads. On the one hand, there is increasing international focus on climate mitigation and energy policy. The global climate movement is making headway to limit the flow and consumption of fossil fuels, and among publics worldwide there is greater awareness of the implications of exporting greenhouse gas emissions. On the other hand, due to recent technological advances, there are more energy resources available than prior to the 2000s, and energy companies are keen to find and maintain consumers of their products. Meanwhile, first constriction and then rebound in energy demand from the COVID-19 pandemic has complicated the global landscape of fossil fuel supply and demand. What will this mean for the communities that host the technologies and infrastructure of the global fossil fuel trade?

Through this volume we have argued that a focus on the *movement* of fossil fuels worldwide is vital for understanding the social landscape of the energy transition. To date, fossil fuel export has largely been left out of energy social science, yet new export projects are impacting communities around the world. Communities are facing similar considerations as those hosting extraction activities—disruptions to place attachment, local environmental impacts, economic impacts, concerns about local autonomy, etc.—but there are some ways that export-based proposals are unique. Export activities connect communities to the global trade of fossil fuels, and some communities are starting to recognize the power they have to disrupt this cycle. Through greater attention to fossil fuel export, energy social science will gain a more comprehensive understanding of the implications of fossil fuel production for people and communities affected far beyond the well or mine.

References

Baalisampang, T., Abbassi, R., Garaniya, V., Khan, F., & Dadashzadeh, M. (2019). Modelling an integrated impact of fire, explosion and combustion products during transitional events caused by an accidental release of LNG. *Process Safety and Environmental Protection, 128*, 259–272. https://doi.org/10.1016/j.psep.2019.06.005

Bomberg, E. (2017). Shale we drill? Discourse dynamics in UK fracking debates. *Journal of Environmental Policy and Planning, 19*(1), 72–88. https://doi.org/10.1080/1523908X.2015.1053111

Clarke, C. E., Budgen, D., Evensen, D. T. N., Stedman, R. C., Boudet, H. S., & Jacquet, J. B. (2019). *Communicating about climate change, natural gas development, and "fracking": U.S. and International perspectives*. Oxford Research Encyclopedia of Climate Science. https://oxfordre.com/climatescience/view/10.1093/acrefore/9780190228620.001.0001/acrefore-9780190228620-e-443.

Delborne, J. A., Hasala, D., Wigner, A., & Kinchy, A. (2020). Dueling metaphors, fueling futures: "Bridge fuel" visions of coal and natural gas in the United States. *Energy Research & Social Science, 61*, 101350. https://doi.org/10.1016/j.erss.2019.101350

Englund, W. (2021). *Engineers raise alarms over the risk of major explosions at LNG plants*. Washington Post. https://www.washingtonpost.com/business/2021/06/03/lng-export-explosion-vce/.

Erickson, P., Lazarus, M., & Piggot, G. (2018). Limiting fossil fuel production as the next big step in climate policy. *Nature Climate Change, 8*(12), 1037–1043. https://doi.org/10.1038/s41558-018-0337-0

Fawcett, M. (2021). *Here comes the death of LNG*. Canada's National Observer. https://www.nationalobserver.com/2021/07/23/opinion/death-of-lng.

Freudenburg, W. R. (1992). Addictive economies: Extractive industries and vulnerable localities in a changing world economy. *Rural Sociology, 57*(3), 305–332. https://doi.org/10.1111/j.1549-0831.1992.tb00467.x

Giordono, L. S., Boudet, H. S., Karmazina, A., Taylor, C. L., & Steel, B. S. (2018). Opposition "overblown"? Community response to wind energy siting in the western United States. *Energy Research & Social Science, 43*, 119–131. https://doi.org/10.1016/j.erss.2018.05.016

Groeger, L. V. (2012). *Pipelines explained: How safe are America's 2.5 million miles of pipelines?* ProPublica. https://www.propublica.org/article/pipelines-explained-how-safe-are-americas-2.5-million-miles-of-pipelines.

Hazboun, Olson, S., & Boudet, H. S. (2020). Public preferences in a shifting energy future: Comparing public views of eight energy sources in North America's Pacific Northwest. *Energies, 13*(8), 1940. https://doi.org/10.3390/en13081940

Hazboun, Olson, S., & Boudet, H. S. (2021). Natural gas – friend or foe of the environment? Evaluating the framing contest over natural gas through a public opinion survey in the Pacific northwest. *Environmental Sociology, 0*(0), 1–14. https://doi.org/10.1080/23251042.2021.1904535

Jaffe, D., Putz, J., Hof, G., Hof, G., Hee, J., Lommers-Johnson, D. A., Gabela, F., Fry, J. L., Ayres, B., Kelp, M., & Minsk, M. (2015). Diesel particulate matter and coal dust from trains in the Columbia River Gorge, Washington State, USA. *Atmospheric Pollution Research, 6*(6), 946–952. https://doi.org/10.1016/j.apr.2015.04.004

Leber, R. (2021). *Gaslit: How the fossil fuel industry convinced Americans to love gas stoves.* Mother Jones. https://www.motherjones.com/environment/2021/06/how-the-fossil-fuel-industry-convinced-americans-to-love-gas-stoves/.

LINGO. (2021). *Leave fossil fuels in the ground.* LINGO. https://www.leave-it-in-the-ground.org/.

Masson-Delmotte, V., Zhai, P., Pirani, A., Connors, S. L., Péan, C., Berger, S., Caud, N., Chen, Y., Goldfarb, L., Gomis, M. I., Huang, M., Leitzell, K., Lonnoy, E., Matthews, J. B. R., Maycock, T. K., Waterfield, T., Yelekçi, Ö., Yu, R., & Zhou, B. (2021). Summary for policymakers. In *Climate change 2021: The physical science basis. Contribution of working group I to the sixth assessment report of the Intergovernmental panel on climate change.* Cambridge University Press.

Mokhatab, S., Mak, J. Y., Valappil, J. V., & Wood, D. A. (2014). Chapter 9 - LNG safety and security aspects. In *Handbook of liquefied natural gas* (pp. 359–435). Gulf Professional Publishing. https://www.sciencedirect.com/science/article/pii/B978012404585900009X.

Pandey, S., Gautam, R., Houweling, S., Gon, H. D. van der, Sadavarte, P., Borsdorff, T., Hasekamp, O., Landgraf, J., Tol, P., Kempen, T. van, Hoogeveen, R., Hees, R. van, Hamburg, S. P., Maasakkers, J. D., & Aben, I. (2019). Satellite observations reveal extreme methane leakage from a natural gas well blowout. *Proceedings of the National Academy of Sciences, 116*(52), 26376–26381. https://doi.org/10.1073/pnas.1908712116

Perry, E. E., Needham, M. D., Cramer, L. A., & Rosenberger, R. S. (2014). Coastal resident knowledge of new marine reserves in Oregon: The impact of proximity and attachment. *Ocean & Coastal Management, 95*, 107–116. https://doi.org/10.1016/j.ocecoaman.2014.04.011

Pierce, J. J., Boudet, H., Zanocco, C., & Hillyard, M. (2018). Analyzing the factors that influence U.S. public support for exporting natural gas. *Energy Policy, 120*, 666–674. https://doi.org/10.1016/j.enpol.2018.05.066

Piggot, G. (2018). The influence of social movements on policies that constrain fossil fuel supply. *Climate Policy, 18*(7), 942–954.

Profita, C. (2019). *Portland city council readopts fossil fuel restrictions.* opb. https://www.opb.org/news/article/portland-city-council-readopts-fossil-fuel-restrictions/.

Rosa, E. A., Machlis, G. E., & Keating, K. M. (1988). Energy and society. *Annual Review of Sociology, 14*(1), 149–172. https://doi.org/10.1146/annurev.so.14.080188.001053

Sightline Institute. (2018). *The thin green line.* Sightline Institute. https://www.sightline.org/research/thin-green-line/.

Sovacool, B. K. (2014). What are we doing here? Analyzing fifteen years of energy scholarship and proposing a social science research agenda. *Energy Research & Social Science, 1*, 1–29. https://doi.org/10.1016/j.erss.2014.02.003

Sovacool, B. K., Axsen, J., & Sorrell, S. (2018). Promoting novelty, rigor, and style in energy social science: Towards codes of practice for appropriate methods and research design. *Energy Research & Social Science, 45*, 12–42. https://doi.org/10.1016/j.erss.2018.07.007

Tonge, J., Ryan, M. M., Moore, S. A., & Beckley, L. E. (2015). The effect of place attachment on pro-environment behavioral intentions of visitors to coastal natural area tourist destinations. *Journal of Travel Research, 54*(6), 730–743. https://doi.org/10.1177/0047287514533010

Weller, Z. D., Hamburg, S. P., & von Fischer, J. C. (2020). A national estimate of methane leakage from pipeline mains in natural gas local distribution systems. *Environmental Science & Technology, 54*(14), 8958–8967. https://doi.org/10.1021/acs.est.0c00437

Woodside. (2021). *Quebec rejects $14B LNG project over environmental concerns.* Canada's National Observer. https://www.nationalobserver.com/2021/07/21/news/climate-advocates-push-quebec-reject-14-billion-lng-project.

Further reading

Hazboun, S. O., Howe, P. D., Layne Coppock, D., & Givens, J. E. (2020). The politics of decarbonization: Examining conservative partisanship and differential support for climate change science and renewable energy in Utah. *Energy Research & Social Science, 70*, 101769. https://doi.org/10.1016/j.erss.2020.101769

Weible, C. M., Heikkila, T., & Carter, D. P. (2017). An institutional and opinion analysis of Colorado's hydraulic fracturing disclosure policy. *Journal of Environmental Policy and Planning, 19*(2), 115–134. https://doi.org/10.1080/1523908X.2016.1150776

Index

Note: 'Page numbers followed by "f" indicate figures those followed by "t" indicate tables and "b" indicate boxes.'

O

Q

R

S

T

U

W

Y

Printed in the United States
by Baker & Taylor Publisher Services